Android
应用与项目开发基础

邵欣欣 付丽梅 王洪岩 严凤龙 刘冰月 ⊙ 编著

清华大学出版社
北京

内 容 简 介

本书以 CoffeeStore 项目为主线,通过 Android 基础开发、界面开发、数据存储和高级开发四部分介绍 Android 的基础知识和高级应用。本书既注重理论介绍,又强调实际应用,从实用的角度精心设计知识结构及代码实例,并配以大量习题,让读者既能掌握计算机语言知识,又能提高实践能力;最后的项目实战可以让读者全面掌握 Android 知识,提高综合应用能力。

本书通俗易懂,简洁明了,实例丰富,既可以作为高校本专科相关专业学生的课程用书,也可作为自学人员的参考资料。

本书封面贴有清华大学出版社防伪标签,无标签者不得销售。
版权所有,侵权必究。侵权举报电话: 010-62782989 13701121933

图书在版编目(CIP)数据

Android 应用与项目开发基础/邵欣欣等编著. —北京: 清华大学出版社,2018(2019.10重印)
ISBN 978-7-302-49581-9

Ⅰ. ①A… Ⅱ. ①邵… Ⅲ. ①移动终端-应用程序-程序设计 Ⅳ. ①TN929.53

中国版本图书馆 CIP 数据核字(2018)第 027497 号

责任编辑:张 玥
封面设计:傅瑞学
责任校对:白 蕾
责任印制:李红英

出版发行:清华大学出版社
　　　网　　址:http://www.tup.com.cn,http://www.wqbook.com
　　　地　　址:北京清华大学学研大厦 A 座　　　邮　编:100084
　　　社 总 机:010-62770175　　　　　　　　　　邮　购:010-62786544
　　　投稿与读者服务:010-62776969,c-service@tup.tsinghua.edu.cn
　　　质量反馈:010-62772015,zhiliang@tup.tsinghua.edu.cn
　　　课件下载:http://www.tup.com.cn,010-62795954

印 装 者:三河市铭诚印务有限公司
经　　销:全国新华书店
开　　本:185mm×260mm　　印 张:24.5　　彩 插:3　　字　数:566 千字
版　　次:2018 年 5 月第 1 版　　　　　　　　　　　印　次:2019 年 10 月第 3 次印刷
印　　数:2001~2500
定　　价:59.50 元

产品编号:074847-01

前 言
PREFACE

《Android 应用与项目开发基础》根据 Android 课程的能力要求和学生的认知规律精心组织了教材内容。

本书是编写课程组所有教师在移动互联网应用开发课程中多年一线授课及项目开发和实训、实践的结晶。本书以 CoffeeStore 项目为主线,通过 Android 基础开发、界面开发、数据存储和高级开发四部分介绍 Android 的基础知识和高级应用,每个章节都配有项目实战和习题,是一本集理论知识、实验项目和课后习题为一体的综合性图书。本书从工程实践的理念出发,以一个课程项目贯穿始终,全面讲述了 Android 的基础知识和核心技术。本书经过作者的精心设计,并配以大量案例和习题,案例既能阐明原理和方法,又具有一定的实用性。本书融教、学、练三者于一体,适合"项目驱动、案例教学、理论实践一体化"的教学模式。

本书编写组成员在移动互联网应用开发领域有丰富的开发和教学经验。近几年指导学生参加多项移动互联网开发领域的比赛,开展大学生创新创业项目,都取得了较好的成绩,且项目组成员与公司合作开发的 APP 项目已上线推广使用。本书的编写充分发挥了各位教师所长,第 1~4、11 章由付丽梅编写,第 5~7 章由邵欣欣编写,第 8~10 章由严凤龙编写,第 12 章由刘冰月编写,第 13~15 章由王洪岩编写,全书最后由邵欣欣和付丽梅统一修改定稿。书中所有例题及相关代码都已在 Android Studio 开发环境中测试通过。

本书的基本结构与内容组织如下。

1. 基本结构

本书共分 4 篇,15 章,以 CoffeeStore App 的项目构思、设计、实施和运行贯穿始终。内容涵盖 Android 应用程序的基本工作原理、Android 界面技术、组件技术、本地存储技术、网络存储技术、服务与广播、定位与地图等多方面的知识。既强调理论,又重视应用。

本书的章节组织如下页所示。

2. 内容组织

本书以项目为导引组织教材内容,下面详细介绍篇和章的内容。

第 1 篇 开发准备——Android 基础开发篇

第 1 章 初识 Android 平台:介绍 Android 开发平台的基本概念、版本发展历程以及系统架构。

第 2 章 搭建 Android 开发与测试环境:开发环境的安装及模拟器的创建。

第 3 章 第一个 Android 应用程序:Android 程序的基本结构、Android 四大组件以及 Activity 的生命周期和不同 Activity 之间的传值。

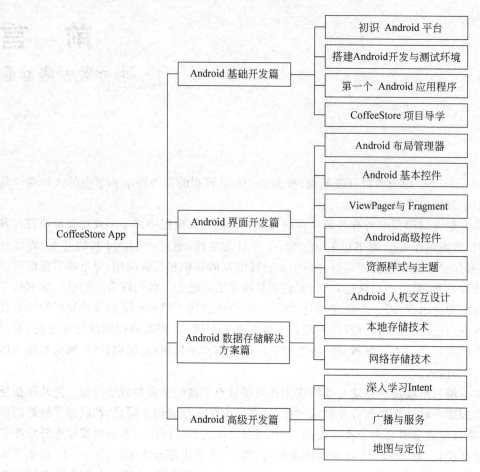

第 4 章 CoffeeStore 项目导学：讲解课程项目 CoffeeStore 的功能需求、体系结构与原型设计和数据库设计。

第 2 篇 界面开发——Android 界面开发篇

第 5 章 Android 布局管理器：线性、相对、表格、网格、帧等常用布局管理器的用法，以及如何向容器中手动添加控件。

第 6 章 Android 基本控件：文本类、按钮类、日期和时间类、进度条、滑动条控件以及星级控件的用法。

第 7 章 ViewPager 与 Fragment：ViewPager 与 PagerAdapter 的用法、Fragment 和 Intent 的用法、Activity 与 Fragment 之间的交互。

第 8 章 Android 高级控件：Adapter 对象、Spinner、Listview、ExpandableListView、GridView 以及 HorizontalScrollView 等高级控件的用法。

第 9 章 资源样式与主题：值资源、位图和色图资源、XML 资源、菜单资源、对话框资源、动画资源、风格资源与主题的用法。

第 10 章 Android 人机交互设计：Android 常用事件、拖拉与多点触屏、手势识别的实现。

第 3 篇　Android 数据存储解决方案篇

第 11 章　本地存储技术：简单数据存储类、Android 文件以及 SQLite 数据库的应用。

第 12 章　网络存储技术：异步任务类、JSON 数据解析以及 HttpURLConnection 的应用。

第 4 篇　Android 高级开发篇

第 13 章　深入学习 Intent：使用 Intent 与 PendingIntent 实现发短信、打电话及系统通知的功能。

第 14 章　广播与服务：广播的基本概念、广播的实现方式，服务的基本概念、服务的生命周期等知识。

第 15 章　地图与定位：地图的定义与显示、地图的定位及路线规划的实现。

由于编者的水平和时间有限，本书的错误和不足在所难免，恳请同行专家和广大读者批评指正。

编　者

2018 年 1 月于大连

目录

第 1 篇 开发准备——Android 基础开发篇

第 1 章 初识 Android 平台 ·················· 3
1.1 Android 简介 ································ 3
1.2 Android 的版本发展历程 ··············· 4
1.3 Android 的制胜法宝 ····················· 5
1.4 Android 的系统架构 ····················· 7
本章小结 ·· 8
本章习题 ·· 8

第 2 章 搭建 Android 开发与测试环境 ····· 9
2.1 安装 Android 开发环境 ················ 9
    2.1.1 安装 JDK ························· 10
    2.1.2 下载和安装 Android Studio 与 Android SDK ········· 10
    2.1.3 Android Studio 开发环境的使用 ···························· 14
2.2 配置 Android 测试环境 ··············· 16
    2.2.1 使用 Android 模拟器运行 Android 程序 ················· 16
    2.2.2 使用真机运行 Android 程序 ···· 19
本章小结 ·· 19
本章练习 ·· 20

第 3 章 第一个 Android 应用程序 ········· 21
3.1 第一个 Android 程序：HelloWorld ··· 21
3.2 Android 程序结构 ······················ 24
3.3 Android 四大组件 ······················ 26
3.4 Activity ···································· 27
    3.4.1 创建和使用 Activity ············ 27

  3.4.2 Activity 的生命周期 ……………………………………………………… 28

  3.4.3 初识 Intent：在不同 Activity 之间传递数据 …………………………… 35

3.5 知识拓展：Activity/ActionBarActivity/AppCompatActivity ………………………… 37

本章小结 …………………………………………………………………………………… 37

本章练习 …………………………………………………………………………………… 37

第 4 章 CoffeeStore 项目导学 ……………………………………………………… 38

4.1 功能描述 ……………………………………………………………………………… 38

4.2 体系结构与知识点 …………………………………………………………………… 39

4.3 原型设计 ……………………………………………………………………………… 41

4.4 数据库设计 …………………………………………………………………………… 42

本章小结 …………………………………………………………………………………… 44

本章习题 …………………………………………………………………………………… 44

第 2 篇 界面开发——Android 界面开发篇

第 5 章 Android 布局管理器 ……………………………………………………… 47

5.1 线性布局管理器 ……………………………………………………………………… 48

5.2 相对布局管理器 ……………………………………………………………………… 53

5.3 表格布局管理器 ……………………………………………………………………… 58

5.4 网格布局管理器 ……………………………………………………………………… 61

5.5 帧布局管理器 ………………………………………………………………………… 63

5.6 向容器中手动添加控件 ……………………………………………………………… 67

5.7 项目实战：CoffeeStore 首页的界面开发 …………………………………………… 69

  5.7.1 项目分析 ………………………………………………………………………… 69

  5.7.2 项目实现 ………………………………………………………………………… 70

  5.7.3 项目说明 ………………………………………………………………………… 76

本章小结 …………………………………………………………………………………… 77

本章习题 …………………………………………………………………………………… 78

第 6 章 Android 基本控件 ………………………………………………………… 79

6.1 文本类控件 …………………………………………………………………………… 79

  6.1.1 TextView ………………………………………………………………………… 79

  6.1.2 AutoCompleteTextView ………………………………………………………… 84

  6.1.3 MultiAutoCompleTextView …………………………………………………… 84

  6.1.4 EditText ………………………………………………………………………… 84

6.2 ScrollView …………………………………………………………………………… 88

6.3 按钮类控件 …………………………………………………………………………… 90

	6.3.1 Button ·· 90
	6.3.2 ImageButton ··· 91
	6.3.3 ToggleButton ·· 94
	6.3.4 CheckBox ··· 94
	6.3.5 RadioButton ·· 95
6.4	日期和时间类控件 ··· 100
	6.4.1 DatePicker ·· 100
	6.4.2 TimePicker ·· 104
	6.4.3 DigitalClock ·· 108
	6.4.4 Chronometer ··· 108
6.5	进度条控件 ProgressBar ··· 112
6.6	滑动条 SeekBar ··· 112
6.7	星级控件 RatingBar ·· 113
6.8	项目实战：使用 RadioButton 实现主页底端导航条 ································· 123
	6.8.1 项目分析 ·· 123
	6.8.2 项目实现 ·· 123
	6.8.3 项目说明 ·· 125
6.9	知识扩展：创建和使用自定义控件 ··· 126
本章小结 ·· 128
本章习题 ·· 128

第 7 章 ViewPager 与 Fragment ·· 130

7.1	ViewPager 与 PagerAdapter ·· 130
7.2	Fragment 及其应用场合 ··· 135
7.3	创建 Fragment ·· 137
7.4	初识 Intent ·· 143
	7.4.1 Intent 对象的基本概念 ··· 143
	7.4.2 Intent 对象的基本使用方法 ·· 144
	7.4.3 使用 Intent 对象在 Activity 之间传递数据 ·································· 146
7.5	Activity 与 Fragment 之间的交互 ··· 149
	7.5.1 为 Activity 创建事件回调方法 ··· 149
	7.5.2 添加项目到 ActionBar ·· 150
	7.5.3 与 Activity 生命周期的协调工作 ··· 151
7.6	项目实战：CoffeeStore 主页滑动功能的实现 ······································ 151
	7.6.1 项目分析 ·· 151
	7.6.2 项目实现 ·· 151
	7.6.3 项目说明 ·· 162
本章小结 ·· 163

本章习题 ………………………………………………………………………… 163

第 8 章　Android 高级控件 …………………………………………………… 164
8.1　Adapter 对象 ……………………………………………………………… 164
8.2　Spinner 控件 ……………………………………………………………… 165
8.3　ListView 控件 …………………………………………………………… 167
8.4　ExpandableListView 控件 ……………………………………………… 182
8.5　GridView 控件 …………………………………………………………… 186
8.6　HorizontalScrollView 控件 …………………………………………… 188
8.7　项目实战：CoffeeStore 首页广告轮播效果 ………………………… 190
　　8.7.1　项目分析 ………………………………………………………… 190
　　8.7.2　项目实现 ………………………………………………………… 190
　　8.7.3　项目说明 ………………………………………………………… 191
8.8　项目实战：CoffeeStore 店铺列表页 ………………………………… 192
　　8.8.1　项目分析 ………………………………………………………… 192
　　8.8.2　项目实现 ………………………………………………………… 192
　　8.8.3　项目说明 ………………………………………………………… 195
8.9　项目实战：CoffeeStore 首页推荐商品 ……………………………… 195
　　8.9.1　项目分析 ………………………………………………………… 195
　　8.9.2　项目实现 ………………………………………………………… 196
　　8.9.3　项目说明 ………………………………………………………… 197
本章小结 ………………………………………………………………………… 198
本章习题 ………………………………………………………………………… 198

第 9 章　资源样式与主题 ……………………………………………………… 200
9.1　资源 ………………………………………………………………………… 200
9.2　值资源 ……………………………………………………………………… 202
　　9.2.1　字符串资源 ……………………………………………………… 202
　　9.2.2　颜色资源 ………………………………………………………… 204
　　9.2.3　尺寸资源 ………………………………………………………… 205
　　9.2.4　数组资源 ………………………………………………………… 206
9.3　位图资源与色图资源 …………………………………………………… 207
9.4　XML 资源 ………………………………………………………………… 208
9.5　菜单资源 ………………………………………………………………… 210
9.6　对话框资源 ……………………………………………………………… 215
　　9.6.1　提醒（Toast）对话框 …………………………………………… 215
　　9.6.2　AlertDialog ……………………………………………………… 218
　　9.6.3　其他对话框资源 ………………………………………………… 223

9.7 动画资源 ……………………………………………………………… 223
9.8 风格资源与主题 ………………………………………………………… 226
  9.8.1 风格资源 …………………………………………………… 226
  9.8.2 主题资源 …………………………………………………… 228
  9.8.3 图像状态资源 ……………………………………………… 230
9.9 国际化(I18N) …………………………………………………………… 231
9.10 项目实战：CoffeeStore 中各种资源的使用 ………………………… 233
  9.10.1 项目分析 ………………………………………………… 233
  9.10.2 项目实现 ………………………………………………… 234
  9.10.3 项目说明 ………………………………………………… 238
本章小结 ………………………………………………………………………… 238
本章习题 ………………………………………………………………………… 239

第 10 章 Android 人机交互设计 …………………………………………… 240

10.1 常用事件 ……………………………………………………………… 240
  10.1.1 按键事件 ………………………………………………… 241
  10.1.2 触摸事件 ………………………………………………… 244
10.2 拖拉与多点触屏 ……………………………………………………… 247
10.3 手势识别 ……………………………………………………………… 248
10.4 项目实战：CoffeeStore 引导页图片切换的实现 …………………… 250
  10.4.1 项目分析 ………………………………………………… 250
  10.4.2 项目实现 ………………………………………………… 250
  10.4.3 项目说明 ………………………………………………… 254
本章小结 ………………………………………………………………………… 254
本章习题 ………………………………………………………………………… 254

第 3 篇   Android 数据存储解决方案篇

第 11 章 本地存储技术 ……………………………………………………… 259

11.1 简单数据存储类 SharedPreferences ………………………………… 259
  11.1.1 SharedPreferences 的使用场合 ………………………… 259
  11.1.2 使用 SharedPreferences 存取数据 ……………………… 260
11.2 Android 文件 ………………………………………………………… 261
  11.2.1 文件数据的存储与读取 ………………………………… 261
  11.2.2 读写 SD 卡中的文件 …………………………………… 267
  11.2.3 读写资源文件 …………………………………………… 271
11.3 SQLite 数据库 ………………………………………………………… 274
  11.3.1 SQLite 数据库存储数据概述 …………………………… 274

11.3.2 使用 SQLiteOpenHelper 类管理数据库版本 ……… 274
11.3.3 使用 SQLiteDatabase 操作数据库 ……… 276
11.3.3 一起发布数据库与应用程序 ……… 279
11.4 项目实战：CoffeeStore 启动页安装信息的存取 ……… 280
 11.4.1 项目分析 ……… 280
 11.4.2 项目实现 ……… 280
 11.4.3 项目说明 ……… 281
11.5 项目实战：读取数据库文件 ……… 281
 11.5.1 项目分析 ……… 281
 11.5.2 项目实现 ……… 281
 11.5.3 项目说明 ……… 284
11.6 项目实战：CoffeeStore 项目中本地收藏夹的实现 ……… 284
 11.6.1 项目分析 ……… 284
 11.6.2 项目实现 ……… 284
 11.6.3 项目说明 ……… 294
本章小结 ……… 295
本章习题 ……… 296

第 12 章 网络存储技术 ……… 297

12.1 异步任务 ……… 297
 12.1.1 异步任务的使用场合 ……… 297
 12.1.2 异步任务类 ……… 298
12.2 JSON 数据解析 ……… 300
 12.2.1 JSON 简介 ……… 300
 12.2.2 JSON 的基本语法 ……… 301
 12.2.3 JSON 的解析 ……… 302
12.3 HttpURLConnection ……… 304
 12.3.1 HTTP 通信接口 ……… 304
 12.3.2 HttpURLConnection 的常用方法 ……… 304
12.4 利用异步任务读取服务器端图片信息 ……… 306
12.5 项目实战：登录功能 ……… 307
 12.5.1 项目分析 ……… 307
 12.5.2 项目实现 ……… 307
 12.5.3 项目说明 ……… 317
12.6 项目实战：店铺列表功能 ……… 317
 12.6.1 项目分析 ……… 317
 12.6.2 项目实现 ……… 318
 12.6.3 项目说明 ……… 321
本章小结 ……… 322

本章习题 ······ 322

第4篇 Android 高级开发篇

第13章 深入学习 Intent ······ 327
13.1 PendingIntent ······ 327
13.2 Intent 过滤器 ······ 328
13.3 运行时权限 ······ 331
本章小结 ······ 345
本章习题 ······ 345

第14章 广播与服务 ······ 346
14.1 广播的定义与用途 ······ 346
14.2 广播接收器的实现 ······ 347
14.3 服务的基本概念 ······ 351
14.4 服务的生命周期 ······ 353
本章小结 ······ 358
本章习题 ······ 358

第15章 地图与定位 ······ 359
15.1 位置服务 ······ 359
15.2 地图的定义与显示 ······ 360
15.2.1 申请地图密钥 ······ 360
15.2.2 地图的显示 ······ 361
15.3 地图的定位及路线规划 ······ 365
15.3.1 定位原理 ······ 365
15.3.2 定位与路线规划 ······ 366
本章小结 ······ 374
本章习题 ······ 374

参考文献 ······ 375

附录 A RGB 颜色对照表 ······ 377

开发准备——Android 基础开发篇

完成一个项目,首先要进行构思和设计,构思和设计是实施、运行的基础。只有把目标弄清楚,才能够有效进行后续工作。也就是说,首先要明确要完成的项目是什么,对项目进行分析和整体设计,然后再采用具体的工具和方法实现项目中的每一个环节,并把各个部分组合起来,构成一个能够正常运行的整体。本书以一个典型的电商类 App 课程项目 CoffeeStore 贯穿,通过学习这个项目,读者将了解一个在线商城项目移动端的主要功能及实现技术,从而掌握 Android 应用程序开发的核心技术。本篇为项目准备篇,分 4 章来讲解,主要内容如下。

- 初识 Android 平台
- 搭建 Android 开发与测试环境
- 第一个 Android 应用程序
- CoffeeStore 项目导学

初识 Android 平台

本章概述

通过本章的学习,掌握 Android 移动应用开发的基本概念,了解 Android 版本的发展历程,了解各个 Android 平台的特点,掌握 Android 的系统架构。

学习重点与难点

重点:

(1) Android 的系统架构。

(2) Android 版本的发展历程。

(3) Android 平台的特点。

难点:

Android 的系统架构。

学习建议

读者需要认真阅读、识记 Android 开发平台的特点。对于 Android 的系统架构,可课后延伸阅读,加深理解。

1.1 Android 简介

Android 是一个以 Linux 为基础的半开源操作系统,主要用于移动设备,由谷歌和开放手持设备联盟开发与领导。Android 系统由 Andy Rubin 制作,最初主要支持手机。2005 年 8 月 17 日,Android 被谷歌收购。2007 年 11 月 5 日,谷歌与 84 家硬件制造商、软件开发商及电信营运商组成开放手持设备联盟(Open Handset Alliance),共同研发改良 Android 系统并生产搭载 Android 的智慧型手机,并逐渐拓展到平板电脑及其他领域。随后,谷歌以 Apache 免费开源许可证的授权方式发布了 Android 的源代码。

Android 的主要竞争对手是苹果公司的 IOS 以及 RIM 的 Blackberry OS。2011 年第一季度,Android 在全球的市场份额首次超过塞班系统,跃居全球第一。2012 年 7 月的数据显示,Android 占据全球智能手机操作系统市场 59% 的份额,中国的市场占有率为 76.7%。

Android 的本意是机器人,Android 的标志也是机器人,这就是 Andy 的理想和信念。从星球大战开始,美国便对机器人有了感情,从各种好莱坞大片中我们知道,美国人一直希望有个机器人朋友……所以可以想象,在未来的生活中,Android 不仅仅是移动终端上的智能平台,还会成为所有智能设备的 DNA,帮助所有设备互相理解、互相通信,并可以被人愉快地使用和控制。自然,Android 这个机器人会给人类带来美好的智能生活。

1.2 Android 的版本发展历程

最早版本的 Android 1.0 Beta 发布于 2007 年 11 月 5 日,至今已经发布了多个更新。这些更新版本都在前一个版本的基础上修复了 bug,并添加了前一个版本没有的新功能。Android 操作系统曾有两个预发布的内部版本,它们的代号分别是铁臂阿童木(Astro)和发条机器人(Bender)。2008 年 9 月,谷歌正式发布了 Android 1.0 系统,这也是 Android 系统最早的版本。由于涉及版权问题,从 2009 年 5 月开始,Android 操作系统改用甜点来作为版本代号,这些版本按照大写字母的顺序来命名。下面展示 Android 几个典型版本的发展演变。

(1) Android 1.5 Cupcake(纸杯蛋糕):2009 年 4 月 30 日发布。

主要更新:拍摄/播放影片,并支持上传到 Youtube;支持立体声蓝牙耳机,同时改善自动配对性能;最新的采用 WebKit 技术的浏览器,支持复制/粘贴和页面中搜索;GPS 性能大大提高;提供屏幕虚拟键盘;主屏幕增加音乐播放器和相框 Widgets;应用程序自动随着手机旋转;短信、G-mail、日历、浏览器的用户接口大幅改进,如 G-mail 可以批量删除邮件;相机启动速度加快,拍摄图片可以直接上传到 Picasa;来电照片显示。

(2) Android 1.6 Donut(甜甜圈):2009 年 9 月 15 日发布。

主要更新:重新设计的 Android Market 手势;支持 CDMA 网络;文字转语音(Text-to-Speech)系统;快速搜索框;全新的拍照接口;查看应用程序耗电;支持虚拟私人网络(VPN);支持更多的屏幕分辨率;支持 OpenCore2 媒体引擎;新增面向视觉或听觉困难人群的易用性插件。

(3) Android 2.0/2.0.1/2.1 Eclair(松饼):2009 年 10 月 26 日发布。

主要更新:优化硬件速度;Car Home 程序;支持更多的屏幕分辨率;改良的用户界面;新的浏览器的用户接口和支持 HTML5;新的联系人名单;更好的白色/黑色背景比率;改进谷歌 Maps 3.1.2;支持 Microsoft Exchange;支持内置相机闪光灯;支持数码变焦;改进的虚拟键盘;支持蓝牙 2.1;支持动态桌面的设计。

(4) Android 2.2/2.2.1 Froyo(冻酸奶):2010 年 5 月 20 日发布。

主要更新:整体性能大幅度提升;3G 网络共享功能;Flash 的支持;App2sd 功能;全新的软件商店;更多的 Web 应用 API 接口的开发。

(5) Android 2.3.x Gingerbread(姜饼):2010 年 12 月 7 日发布。

主要更新:增加新的垃圾回收和优化处理事件;原生代码可直接存取输入和感应器事件、EGL/OpenGLES、OpenSL ES;新的管理窗口和生命周期的框架;支持 VP8 和 WebM 视频格式;提供 AAC 和 AMR 宽频编码;提供新的音频效果器;支持前置摄像头、SIP/VOIP 和 NFC(近场通讯);简化界面、速度提升;更快更直观的文字输入;一键文字选择和复制/粘贴;改进的电源管理系统;新的应用管理方式。

(6) Android 3.1 Honeycomb(蜂巢):2011 年 5 月 11 日发布。

主要更新:经过优化的 G-mail 电子邮箱;全面支持谷歌 Maps;将 Android 手机系统跟平板系统再次合并,从而方便开发者;任务管理器可滚动,支持 USB 输入设备(键盘、鼠标等);支持谷歌 TV,可以支持 XBOX 360 无线手柄;Widget 支持的变化,能更加容易地定制屏幕 Widget 插件。

(7) Android 4.2 Jelly Bean(果冻豆):2012 年 10 月 30 日线上发布。

谷歌原定于 2012 年 10 月 30 日召开 Android 4.2 的发布会,但由于受到桑迪(Sandy)飓风的影响而临时取消。不过谷歌仍通过其官方博客发布了全新的 Android 4.2 系统。Android 4.2 沿用了 4.1 版"果冻豆"(Jelly Bean)这一名称,与 Android 4.1 有很高的相似性,但仍在细节上做了一些改进与升级,尤其是在安全性方面进行了提升。

重要更新:Photo Sphere 全景拍照;键盘手势输入;Miracast 无线显示共享;手势放大缩小屏幕,以及为盲人用户设计的语音输出和手势模式导航功能等。令人关注的是,谷歌在 Android 4.2 中新加入了新的恶意软件扫描功能。

(8) Android 5.1 Lollipop(棒棒糖):2014 年 6 月 26 日发布。

谷歌在 2014 年 6 月 26 日的 I/O 2014 开发者大会上正式推出了 Android 5.1,可以说是 Android 系统自 2008 年问世以来变化最大的升级版本。除了新的用户界面、性能升级和跨平台支持,全面的电池寿命增强及更深入的应用程序集成也令人印象深刻。

(9) Android 6.0 Marshmallow(棉花糖):2015 年 9 月 30 日发布。

2015 年 9 月 30 日凌晨,谷歌在美国旧金山举行 2015 年秋季新品发布会。在发布会上,代号为 Marshmallow(棉花糖)的 Android 6.0 系统正式推出。新系统的整体设计风格依然保持扁平化的 MeterialDesign 风格。Android 6.0 在软件体验与运行性能上进行了大幅度的优化。据测试,Android 6.0 可使设备续航时间提升 30%。

(10) Android 7.0 Nougat(牛轧糖):2016 年 8 月 22 日发布。

2016 年到来的 Android 系统也将拥有 iPhone 6s 的 3D Touch 功能。谷歌将会学习 iPhone 6s 上的 3D Touch 功能,为 Android 用户提供相似的移动体验。针对安卓 7.0 的更新版本 Android 7.1 已经发布,增加了 emoji 表情与 GIF 键盘功能,并且采用了全新的圆形图标设计。

(11) Android 8.0 Oreo(奥利奥):2017 年 8 月 22 日发布。

2017 年,Android 8.0 初期仅向"安卓开源计划(Android Open Source Project)"的用户开放,对谷歌的 Pixel 和 Nexus 手机用户,在不久的将来也将开放更新。奥利奥版 Android 的聚焦重点是电池续航能力、速度和安全,让用户更好地控制各种应用程序。

1.3 Android 的制胜法宝

Android 到底有什么魅力,可以让众多的粉丝为之疯狂?据粗略统计,Android 至少有如下 7 项优势。

(1) Android 平台是免费、开源的。开源的好处就是人人可以成为内容的创造者,利于创新,不足之处就是会有恶意软件混进去。Android OS 采取开源的推广方式,任何厂商都可以免费使用 Android OS 作为其操作系统软件。因此，Android 手机种类异常丰富，比 iPhone 更具价格上的优势，因此市场占有率持续攀升。

另外，作为开源的操作系统，Android 的软件也更丰富，开发潜力更大。近几年，国产手机厂商如小米、魅族也通过 Android OS 定制出优秀的 MIUI FlymeOS 等系统，已经完全能够满足人们的日常需求。

还有，Android OS 可以应用在其他诸如导航仪、数码播放器甚至数码电视这样的设备上，应用范围非常广。

此外，自 Android 4.1.1 版本以来，Android OS 的性能日趋完善，配置稍高的 Android 手机运行的流畅度已经和 iPhone 不相上下。反观 iPhone，IOS 8.1 更新之后，反而出现了之前版本从未有过的卡顿问题，IOS 的通知栏也有了 Android 的影子，可见近几年 IOS 的新功能开发略显疲态，若不是中国、欧美等主要市场一大批忠实用户的簇拥，如今的 IOS 已经难以与 Android OS 相抗衡。至于 Windows Phone OS 和 Blackberry OS 等系统，虽然也有着一定的市场占有率，但也难以撼动 Android OS 在智能手机市场上绝对的统治力，已经逐渐变成了小众的操作系统。

（2）我的平台我做主。Android 上的所有应用程序都是可替换和扩展的，即使是拨号、Home 这样的核心组件也一样。只要有足够的想象力，就可以缔造出一个独一无二、完全属于自己的 Android 世界。

（3）拥抱 Web 时代。如果想在 Android 应用程序中嵌入 HTML、JavaScript，真是再容易不过了。基于 Webkit 内核的 WebView 组件会完成一切。更值得一提的是，JavaScript 还可以和 Java 无缝整合在一起。互联网巨头谷歌推动的 Android 终端天生就有网络特色，将让用户离互联网更近。

（4）丰富的硬件选择。这一点还是与 Android 平台的开放性相关，由于 Android 具有开放性，众多厂商会推出千奇百怪、特色功能多样的多种产品。功能上的差异和特色，却不会影响到数据同步及软件的兼容，好比从诺基亚 Symbian 风格手机一下改用苹果 iPhone，同时还可将 Symbian 中优秀的软件带到 iPhone 上使用，联系人等资料更是可以方便地转移，是不是非常方便呢？

（5）不受任何限制的开发商。Android 平台给第三方开发商提供一个十分宽泛、自由的环境，不会受到各种条条框框的阻挠，可想而知，会有多少新颖别致的软件诞生！但其也有两面性，血腥、暴力、情色方面的程序和游戏的控制正是留给 Android 的难题之一。

（6）无缝结合的谷歌应用。如今叱咤互联网的谷歌已经走过 10 年历史，从搜索巨人到全面的互联网渗透，谷歌服务如地图、邮件、搜索等已经成为连接用户和互联网的重要纽带，而 Android 平台手机将无缝结合这些优秀的谷歌服务。

（7）个性的充分体现。21 世纪是崇尚个性的时代。Android 也紧随时代潮流，提供了众多体现个性的功能。Widget、Shortcut、Live WallPapers，无一不尽显手机的华丽与时尚。

1.4　Android 的系统架构

图 1.1 从技术层面描述了 Android 系统架构中各层的关系。

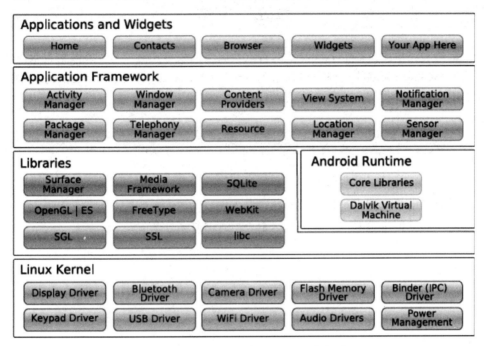

图 1.1　Android 系统架构中的层次关系

从图 1.1 可以看出，Android 系统架构分为 4 层，从低到高分别是 Linux 内核层、系统运行库、应用程序框架和应用程序。下面对这 4 层进行简单的介绍。

Linux 内核：第一层相当于硬件层和软件层之间的一个抽象层，提供显示驱动、摄像头驱动、闪存驱动、键盘驱动、音频驱动、电源管理、Wi-Fi 驱动等功能，使得 Android 能实现核心系统服务。

系统运行库：可以分成两部分，分别是系统库和 Android 运行时，系统库是应用程序框架的支撑，是连接应用程序框架层与 Linux 内核层的重要纽带。Android 应用程序时采用 Java 语言编写，程序在 Android 运行时中执行，运行时分为核心库和 Dalvik 虚拟机两部分。核心库提供了 Java 语言 API 中的大多数功能，同时也包含了 Android 的一些核心 API。Dalvik 虚拟机是一种基于寄存器的 Java 虚拟机，而不是传统的基于栈的虚拟机，并进行了内存资源使用的优化，具有支持多个虚拟机的特点。

应用程序框架：第三层是 Android 核心应用程序所使用的 API 框架，是创建应用程序时需要使用的各种高级构建块。用户可以自由地使用它们来开发自己的应用程序。该框架最重要的部分包括活动管理器、内容提供器、资源管理器、位置管理器、通知管理器。

应用程序：Android 平台不仅仅是操作系统，也包含了许多应用程序，诸如 SMS 短信客户端程序、电话拨号程序、图片浏览器、Web 浏览器等应用程序。这些应用程序都是用 Java 语言编写的,并且都可以被开发人员开发的其他应用程序所替换,这点不同于其他手机操作系统固化在系统内部的系统软件,因此更加灵活和个性化。

本章小结

本章从宏观角度介绍了 Android 的特点及版本的发展历程。Android 的发展基本上就是从丑小鸭到白天鹅的过程。在未来的 3～5 年内,Android 还会得到怎样的发展呢?让我们一起憧憬着 Android 的美好未来!

本章习题

1. Android 经历了哪些版本?
2. Android 平台的主要特点有哪些?
3. Android 的系统架构包括哪几层?各层都有什么作用?

搭建 Android 开发与测试环境

本章概述

通过本章的学习,掌握 Android 开发环境的搭建步骤,这是学习 Android 开发的第一步。有了开发环境还无法测试 Android 程序,还需要配置 Android 的测试环境,包括模拟器和真机测试环境。

学习重点与难点

重点:

(1) 搭建 Android 开发环境。

(2) Android 开发环境的使用。

(3) 用模拟器测试 Android 程序。

(4) 用真机测试 Android 程序。

难点:

用真机测试 Android 程序。

学习建议

读者需要不断实践搭建 Android 开发与测试环境。由于操作系统不同、机型不同,安装过程中可能会遇到各种各样的问题,读者要耐心分析问题,通过阅读本章和利用网络资源来解决安装过程中的问题。

2.1 安装 Android 开发环境

谷歌公司推荐的最新 Android 开发环境是 Android Studio,它是谷歌官方在 2013 年谷歌 I/O 大会上发布的全新 Android 开发 IDE,基于 IntelliJ IDEA 开发环境。Android Studio 提供了集成的 Android 开发工具,用于开发和调试。Android 开发环境必需的工具如下。

- JDK(7 及以上版本)
- Android Studio
- Android SDK

2.1.1 安装 JDK

最新版本的 JDK 可到 Oracle 公司的官网上下载,本书下载的是 JDK7 的版本。

下载到本地电脑后双击图标进行安装。JDK 的安装过程比较简单,安装的时候注意将 JDK 和 JRE 安装到同一个目录即可,JDK 默认安装成功后,系统目录下会出现两个文件夹,一个代表 JDK,一个代表 JRE。

安装注意事项如下。

(1) Android Studio 要求 JDK 版本为 JDK7 及更高。

(2) 确认电脑操作系统是 32 位还是 64 位,一定下载对应的 JDK 版本。Windows x86 对应 Windows 32 位机器,Windows x64 对应 Windows 64 位机器。如果安装的 Android Studio 与 JDK 不匹配,打开时会报错。

(3) JDK 的环境变量一定按链接中的要求配置好,使用传统的 JAVA_HOME 环境变量名称,否则打开 Android Studio 时会因为找不到 JDK 的路径报错。

2.1.2 下载和安装 Android Studio 与 Android SDK

下载之前,要确保已经安装 JDK。国内不能直接下载,但有两种下载方式可供选择,一种是官网,另一种是国内的镜像网站。

下载完后,如果是安装包,直接安装即可;如果是解压包,可以解压后直接运行。安装目录为/studio.exe 文件。打开之后,会进入设置页面,如果没有安装 SDK,选择 Standard(标准),如果已经安装了 SDK,就选择 Custom(自定义),然后选择 SDK 安装目录即可。

(1) 无论是下载压缩包还是 exe 文件,双击进入即可。

android-studio-bundle-135.1740770-windows

(2) 进入后按照向导指示安装,首先出现图 2.1 所示的欢迎对话框。

图 2.1 欢迎对话框

(3) 进入图 2.2 所示的对话框,第 1 个是 Android Studio 主程序,必选。第 2 个是 Android SDK,若要安装 SDK,也选中,若已有 SDK,可不选。第 3 个和第 4 个是虚拟机和虚拟机的加速程序,如果要使用虚拟机调试程序,就选中。完成后单击 Next 按钮,进入图 2.3 所示对话框。

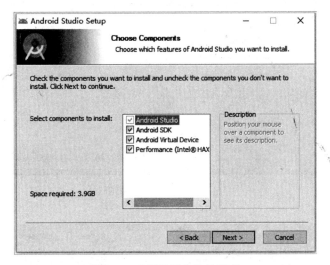

图 2.2 选择需要安装的选项

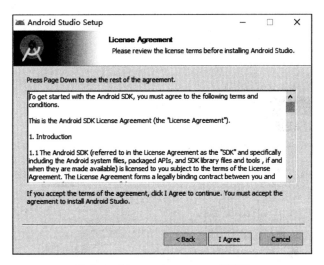

图 2.3 License Agreement 对话框

(4) 若事先有 SDK,则可导入 SDK,如图 2.4 所示。

(5) 若没有 SDK,则需要在线安装,如图 2.5 所示。注意可能因为谷歌网站上不去而不能在线安装的问题。

(6) 选择 Android Studio 和 SDK 的安装目录,如图 2.6 所示。
　　选择习惯安装软件的磁盘,下面的 SDK 路径,配置时还用得上。

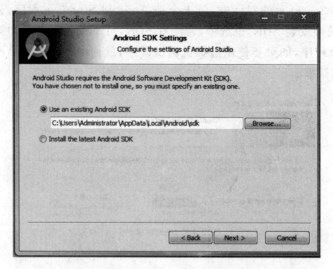

图 2.4 导入现有 SDK

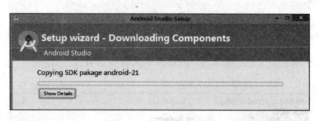

图 2.5 在线安装 SDK

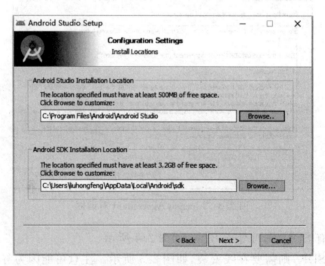

图 2.6 选择安装路径

(7) 设置虚拟机硬件加速器可使用的最大内存,如图 2.7 所示。

如果电脑配置还不错,默认设置 2GB 即可,如果配置比较差,选 1GB 就可以,过大也会影响运行其他软件。

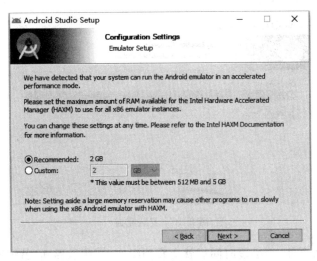

图 2.7　设置可以使用的内存

（8）单击 Next 按钮，进入自动安装模式，如图 2.8～图 2.10 所示。

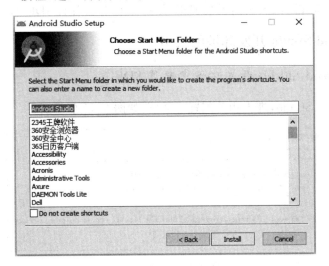

图 2.8　选择安装路径

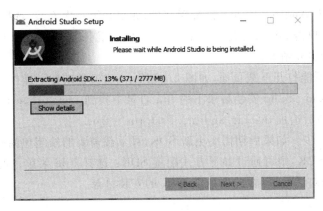

图 2.9　开始安装

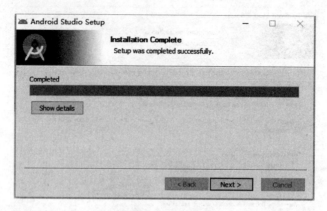

图 2.10　安装完成

如果没出什么意外，一小段时间后就会看到图 2.10 所示界面，就说明安装成功了。安装注意事项如下。

- 安装目录中不允许出现中文。
- Android Studio 的安装前提是 JDK 安装成功，务必保证 JDK 与 SDK 安装成功。

2.1.3　Android Studio 开发环境的使用

1. 在 Android Studio 开发环境中配置 SDK

打开 Android Studio，进入相关配置对话框，如图 2.11 所示。

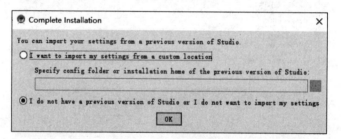

图 2.11　配置对话框

这是用于导入 Android Studio 的配置文件，如果是第一次安装，选择最后一项：不导入配置文件，然后单击 OK 按钮即可。上一步完成后，就会进入图 2.12 所示页面，这是程序在检查 SDK 的更新情况。

如果计算机不能打开下载页面，如图 2.13 所示，建议尝试如下操作。

（1）在 Android Studio 安装目录下的 bin 目录中找到 idea.properties 文件。

（2）在文件最后追加 disable.android.first.run=true。

（3）跳过这一步。如果后期需要更新 SDK，可下载需要的安装包离线配置。

如果已下载 SDK，则可通过如下方式配置 SDK：选择 File 菜单下的 Settings 命令，弹出图 2.14 所示对话框，可在此处指定 SDK 的安装目录。

第 2 章　搭建 Android 开发与测试环境

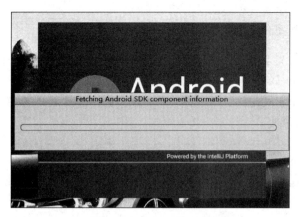

图 2.12　检查 SDK 更新情况

图 2.13　不能打开下载页面

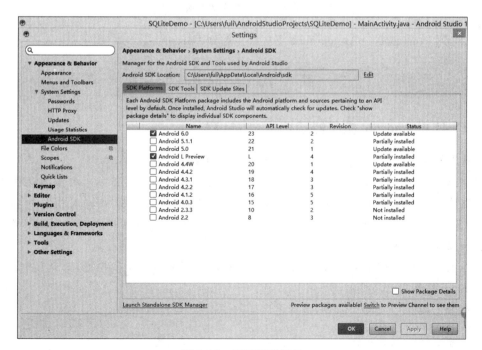

图 2.14　SDK 下载路径

2. 选择主题与字体

（1）同样进入设置界面，单击 IDE Settings 中的 Editor→Colors&Fonts→Font 命令，在 Scheme 中选择 Darcula copy2，就能变成酷炫又护眼的黑色界面。然后单击 Save As...按钮，取一新名字，如图 2.15 所示，并根据预览界面的效果选择一个合适的 Size 数值，即可完成字体的修改。

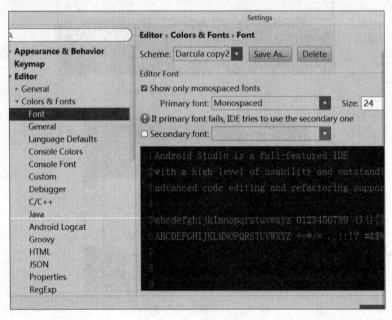

图 2.15　设置字体页面

（2）到 Preference→Appearance 下更改主题到 Darcula。

2.2　配置 Android 测试环境

2.2.1　使用 Android 模拟器运行 Android 程序

首先创建 Android 模拟器。在工具栏上单击 ■ 按钮，启动 Android Device Monitor 工具，再单击 Android Virtual Device Manager ■ 工具，弹出图 2.16 所示模拟器管理对话框。然后单击右侧的 Create 按钮，在弹出的创建模拟器对话框中输入相关内容，如图 2.17 所示。在 Device 列表中选择屏幕分辨率，在 Target 列表中选择 Android 版本，在 Size 文本框中输入 SD Card 的尺寸。

在工具栏上单击 ■ 按钮，出现图 2.18 所示界面，当前有 2 个模拟器，选择其中一个，单击 Start 按钮启动模拟器。

单击工具栏上的 Run 命令，选择已启动的模拟器，运行效果如图 2.19 所示。

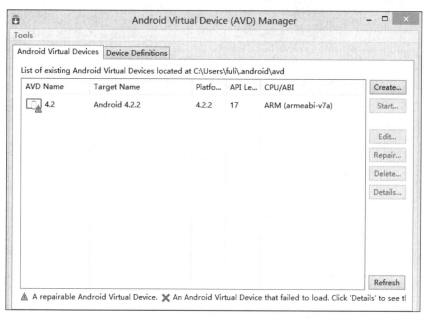

图 2.16　模拟器管理对话框

图 2.17　创建模拟器对话框

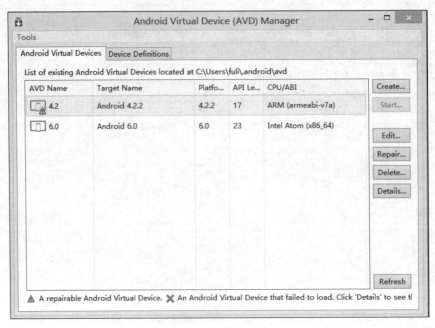

图 2.18 已创建的模拟器

图 2.19 HelloWorld 运行效果图

多数情况下运行 Android Studio，系统会提醒 JDK 或者 Android SDK 不存在，需要重新设置。此时需要到全局的 Project Structure 页面下设置。要进入全局的 Project Structure 页面，在启动页面右下角选择 Configure→Project Defaults→Project Structure 命令。在图 2.20 所示的页面下设置 JDK 或者 Android SDK 目录即可。

第 2 章 搭建 Android 开发与测试环境

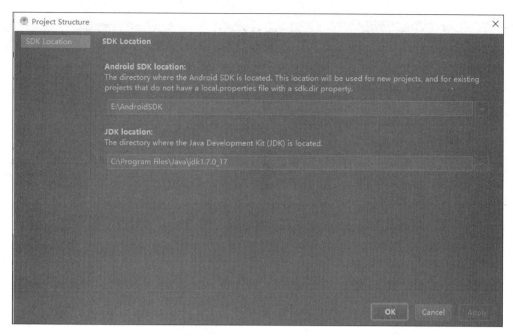

图 2.20 设置 SDK 与 JDK 路径

2.2.2 使用真机运行 Android 程序

尽管在 Android 模拟器中可以测试大多数 Android 应用程序,但是 Android 模拟器运行的速度非常慢(与真机相比),将 APK 安装到 Android 模拟器的过程也非常慢,而且模拟器无法测试使用蓝牙、传感器、NFC 等技术的程序。因此,建议读者尽量使用真机测试。

使用真机测试,首先要用 USB 线将手机与 PC 相连。若是 Ubuntu Linux 和 Mac OS X,在不用安装其他驱动程序的情况下可以检测到几乎所有的 Android 手机和平板电脑。如果是 Windows,基本上检测不到任何 Android 真机设备,所以要为特定的 Android 手机安装相应的驱动程序。很多读者不知道在哪里下载这些驱动程序,建议利用一些手机管理软件自带的驱动程序,让 Windows 识别 Android 手机或者平板电脑。

现在的手机管理软件很多,如 91 助手、360 手机助手等。安装完手机助手后,用 USB 线将手机与 PC 连接,并运行手机助手,第一次运行会安装驱动程序,安装完后,如果手机助手可以检测到手机,那么 Android 开发环境就可以检测到手机。使用真机测试程序,注意需要打开手机的"开发者选项",不同手机的开发者选项,打开的方法略有区别,读者可以查找自己手机打开的方法。

本 章 小 结

搭建 Android 开发和测试环境是 Android 开发的第一步,只有在真实的环境中才能更好地理解和使用环境中的各种工具,积累各种开发技巧。

本 章 练 习

1. 更改 Android Studio 开发环境的主题。
2. 安装 Android Studio 时是否必须选择安装 SDK？为什么？
3. 如何使用真机测试 Android 程序？
4. 如何创建一个模拟器？
5. 尝试安装 Android 开发环境，并记录安装和配置过程中遇到的问题。

第一个Android应用程序

本章概述

本章通过第一个Android程序讲解Android程序的基本结构,Android的组件式开发思想,Android的四大组件,Android中Activity的基本概念、生命周期函数以及Activity之间的传值。

学习重点与难点

重点：

(1) Android程序的基本结构。

(2) Android的四大组件。

(3) Activity的生命周期。

(4) Android的Activity基本用法。

难点：

(1) Android程序的基本结构。

(2) Activity的生命周期。

学习建议

读者要掌握在Android Studio中建立一个Android工程的方法。理解Android程序结构及组件式开发思想,通过反复练习实践掌握Activity的用法,同时通过开发环境的Logcat观察Android生命周期函数的执行时机。

3.1 第一个Android程序：HelloWorld

【例3-1】 创建第一个HelloWorld项目。

第一次启动Android Studio,会出现如图3.1所示的对话框。其中几个选项如下。

选项1：创建一个Android Studio项目。

选项2：打开一个Android Studio项目。

选项3：从版本控制系统中导入代码。支持CVS、SVN、Git、Mercurial甚至GitHub。

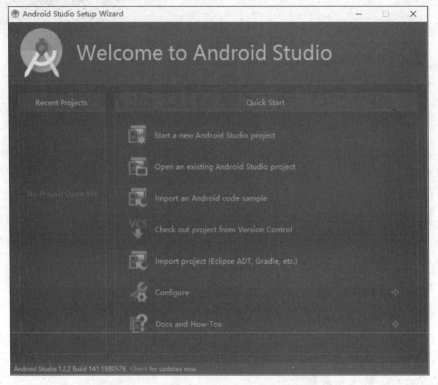

图 3.1　启动对话框

选项 4：导入非 Android Studio 项目，如原生的 Eclipse Android 项目、IDEA Android 项目。如果 Eclipse 项目使用官方建议导出（即使用 Generate Gradle build files 的方式导出），建议使用选项 2 导入。

选项 5：导入官方样例，可从网络下载代码。此功能在以前的测试版本中是没有的，建议多看一看官方的范例。

选项 6：设置。

选项 7：帮助文档。

这里选择第 1 项。

填写应用名和包名，如图 3.2 所示。

选择 Android 版本，如图 3.3 所示。

选择 Activity 的类型，Activity 是 Android 应用程序的基本功能单元，通常用来与用户交互，后续内容会详述，界面如图 3.4 所示。

这里选择一个 Activity 模板，和 Eclipse 很像，直接选择一个 Blank Activity，进入如图 3.5 所示的对话框。

第 3 章 第一个 Android 应用程序

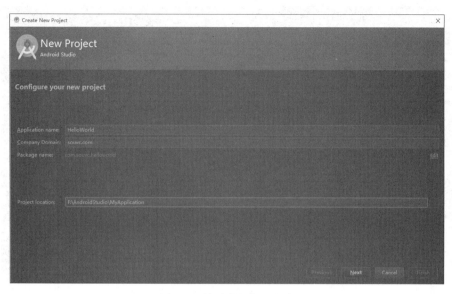

图 3.2 填写应用名和包名

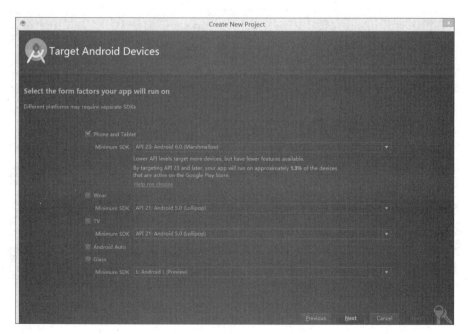

图 3.3 选择 Android 版本

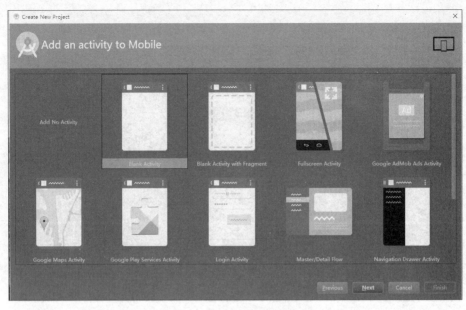

图 3.4　选择 Activity 的类型

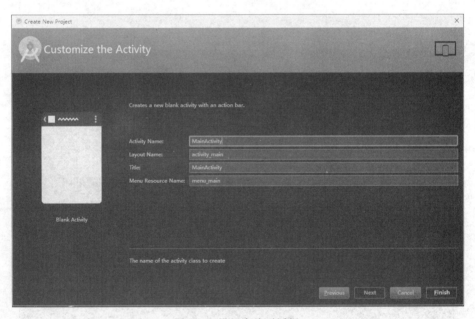

图 3.5　设置完成对话框

3.2　Android 程序结构

Android Studio 提供了图 3.6 所示的几种项目结构类型。

一般常用的有两种结构：Project 结构类型和 Android 结构类型。Project 结构类型如图 3.7 所示。

图 3.6　项目结构类型　　　　图 3.7　Project 结构类型目录结构

各目录功能结构说明如下。
- app/build　app 模块 build 编译输出的目录。
- app/build.gradle　app 模块的 gradle 编译文件。
- app/app.iml　app 模块的配置文件。
- app/proguard-rules.pro　app 模块的 proguard 文件。
- build.gradle 项目的 gradle 编译文件。
- settings.gradle 定义项目包含哪些模块。
- gradlew 编译脚本，可以在命令行执行打包。
- local.properties 配置 SDK/NDK。
- MyApplication.iml 项目的配置文件。
- External Libraries 项目依赖的 Lib，编译时自动下载。

Android 结构类型如图 3.8 所示。

各目录功能结构说明如下。
- app/manifests AndroidManifest.xml 配置文件目录。相当于一个部署文件，指明应用所在的包、应用的图标。对 Activity、Service、Broadcast 进行声明注册后，通过 intent-filter 意图过滤器就能实现组件之间的切换，还需设置应用的权限、版本。
- app/java 源码目录。这里写的是普通 Java 类，它继承于 Activity 的类，即项目源代码。如果工程比较复杂，可以分成许多包，每个包中可以有很多类，这样可以使代码结构清晰，便于开发阅读。
- app/res 资源文件目录。存放应用程序用到的资源文件，包含 anim、drawable、

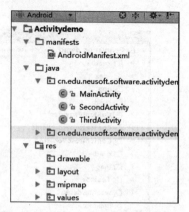

图 3.8　Android 结构类型目录结构

layout、values、raw 等目录。当这个文件下的文件发生变化时，R 会自动发生变化。R 文件在 Package 模式下查看。或者在 Project 模式下依次选择 app→build→generated→source→r→debug 命令，debug 中两个选项的子文件中分别有一个 R 文件即为要找的 R.java 文件。

- Gradle Scripts gradle 编译相关的脚本。

3.3　Android 四大组件

Android 应用开发是组件式开发。所谓组件，可以理解为装修房屋的一个组成元素。一个 Android 应用就像一间房子，房子里的一张桌子、一把椅子、一张床就相当于 Android 的组件。除了这种可以直观看得到的组件，还有一些组件负责服务功能，例如房子中的水管道、煤气管道、电力管道等等。这些组件不是直接呈现在视觉中，但起着很重要的作用，就相当于 Android 的 Service 组件。要实现一个 Android 应用，就可以把一堆接口标准、封装完整的组件拿来用，组件搭配使用，就形成了一间装修好的房子，也就是一个 Android 应用了。

Android 的四大组件如下。

① Activity：表示一个可视化的用户界面，在应用程序中是一个单独的屏幕。每个屏幕都是通过继承和扩展基类 Activity 实现的。

② Service：表示服务，没有可见的用户界面，只提供服务，能够长时间运行于后台，通过继承和扩展基类 Service 来实现。在后台运行于应用程序进程的主线程中，因此 Service 不会阻塞其他组件或者用户界面。

③ ContentProvider：表示内容提供者，可以将应用程序特定的数据提供给另一个应用程序使用，其数据存储方式可以是 Android 文件系统、SQLite 数据库或者其他方式。

④ BroadcastReceiver：表示广播接收器，自身并不实现图形用户界面，但当收到某个广播后，BroadcasetReceiver 可以启动 Activity 作为响应，或者通过 NotificationManager 提醒用户，或者调用 Service 处理长时间事务。

除了以上四大组件外,Intent 也是个非常重要的组件,它在不同组件之间传递信息,将一个组件的请求意图传给另一个组件,可以实现组件之间调用,还可以通过 Intent 进行组件间传递数据。Android 会根据意图的内容选择适当的组件来调用。

3.4　Activity

Activity 的中文意思是活动,可以显示由几个 View 组件组成的用户接口,并且可以对事件进行相应的处理。在 Android 中,Activity 代表手机屏幕的一屏,或是平板电脑中的一个窗口。它是 Android 应用程序与用户交互的窗口,几乎每一个 Android 应用程序都离不开 Activity。它就像一个网站的页面一样,每个页面都可以通过一个独立的类来表示,这个独立的类继承于 Activity 这个基类。

3.4.1　创建和使用 Activity

创建 Activity 大致可以分为以下两个步骤。

(1) 创建一个 Activity,一般是继承 android.app 包中的 Activity 类,不过在不同的应用场景下,也可以继承 Activity 的子类。创建一个继承 Activity 类的 Activity,名称为 MainActivity,具体代码如下。

```
import android.app.Activity;
public class MainActivity extends Activity {
}
```

(2) 重写需要的回调方法。通常情况下,都需要重写 onCreate()方法,并且在该方法中调用 setContentView()方法,设置要显示的视图。例如,在步骤(1)中创建的 Activity 中重写 onCreate()方法,并且设置要显示视图,具体代码如下。

```
@Override
public void onCreate(Bundle savedInstanceState){
    super.onCreate(savedInstanceState);
    setContentView(R.layout.main);
}
```

创建 Activity 后,还需要在 AndroidManifest.xml 文件中配置该 Activity。

具体的配置方法是在＜application＞＜/application＞标记中添加＜activity＞＜/activity＞标记实现。＜activity＞标记的基本格式如下。

```
<activity
    android:icon="@drawable/图标文件名"
    android:name="实现类"
    android:label="说明性文字"
    android:theme="要应用的主题"
    ...
>
```

...
</activity>

启动 Activity 的步骤如下。

(1) 生成一个意图对象 Intent。

(2) 调用 setClass 方法设置所要启动的 Activity。

(3) 调用 startActivity 方法启动 Activity,语句如下。

public void startActivity(Intent intent)

关闭 Activity 的代码如下。

public void finish()

3.4.2 Activity 的生命周期

在 Activity 的生命周期中,有表 3.1 所示的 4 个重要的状态。

表 3.1 Activity 的生命周期

状　　态	描　　述
活动状态	当前 Activity 位于 Activity 栈顶,用户可见,并且可以获得焦点
暂停状态	失去了焦点的 Activity 仍然可见,但是在内存低的情况下,它不能被系统 killed(杀死)
停止状态	该 Activity 被其他 Activity 所覆盖,不可见,但是它仍然保存所有的状态和信息,不过,在内存低的情况下,它将要被系统 killed(杀死)
销毁状态	该 Activity 结束,或 Activity 所在的 Dalvik 进程被结束

图 3.9 描述了 Activity 从创建到销毁整个生命周期方法的调用过程。

从最初调用 onCreate()到最终调用 onDestroy(),称为完整生命周期。Activity 会在 onCreate()进行所有"全局"状态的设置,在 onDestroy()中释放所有持有的资源。例如,如果它有一个从网络上下载数据的后台线程,可能就会在 onCreate()中创建这个线程,并在 onDestroy()中停止这个线程。

从 Activity 调用 onStart()开始,到调用对应的 onStop()为止,称为可见生命周期。在这段时间内,尽管此 Activity 并不一定是在屏幕的最前方,也不一定可以和用户交互,但是用户可以在屏幕上看到这个 Activity。在这两个方法运行周期之间,可以根据需求维护 Activity。例如,可以在 onStart()中注册一个 IntentReceiver(意图接收器)来监控那些可以对 UI 产生影响的环境改变,当 UI 不继续在用户面前显示时,可以在 onStop()中注销这个 IntentReceiver。

每当 Activity 在用户面前显示或者隐藏时,都会调用相应的方法,所以 onStart()和 onStop()方法在整个生命周期中可以多次被调用。

从 Activity 调用 onResume()开始,到调用对应的 onPause()为止,称为前景生命周期。在这段时间内,Activity 处于其他所有 Activity 的前面,且与用户交互。一个 Activity 可以经常在 resumed 和 paused 状态之间转换,例如手机进入休眠时、Activity 的

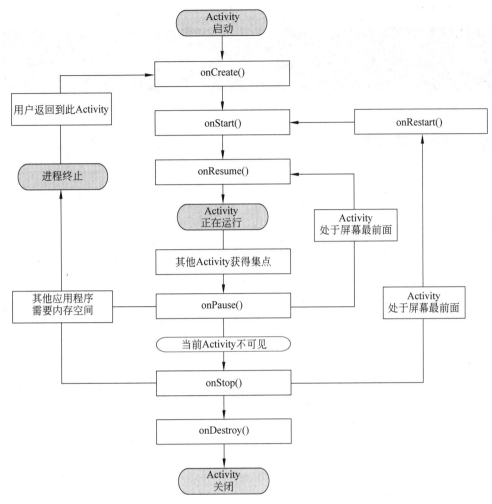

图 3.9 Activity 的生命周期

结果返回时或者新的 Intent 到来时,所以这两个方法中的代码应该非常简短。

总之,所有 Activity 都应该实现自己的 onCreate()方法进行初始化设置,大部分还应该实现 onPause()方法提交数据的修改,并且准备终止与用户的交互。

【例 3-2】 Activity 生命周期回调函数执行时机练习。定义两个 Activity,第一个 Activity 中有个标签,标签的内容为"第一个 Activity",还有一个 Button,显示的内容为 "跳到第二个 Activity"。第二个 Activity 有一个标签,标签内容为"第二个 Activity",还 有一个 Button,显示的内容为"返回第一个 Activity"。在每个 Activity 中添加 Activity 的回调函数,运行观察 Logcat 的输出,理解 Activity 各个回调函数的执行时机。

本程序需要创建两个 Activity,分别为 MainActivity 和 SecondActivity,创建 Activity 的步骤如下。

(1) 使用 Android 项目结构时,选择 Java 目录下的包名,右击,在弹出的快捷菜单中选择 New→Activity→Empty Activity 命令,弹出如图 3.10 所示的对话框。

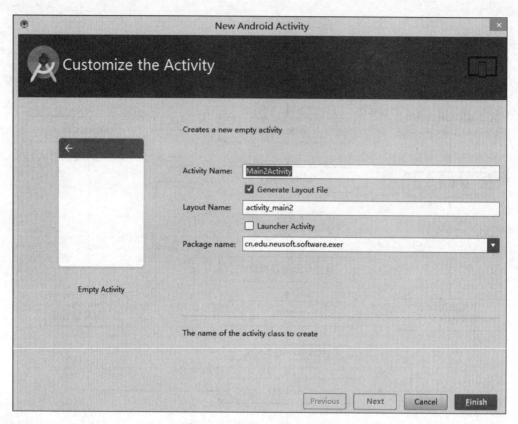

图 3.10 新建 Activity 对话框

（2）在 SecondActivity 类中覆盖 Activity 的几个生命周期方法，在每个方法的方法体里输出当前所在 Activity 名：当前所在方法名。例如，在 onRestart 方法中加入如下输出语句。

```
System.out.println("MainActivity: onRestart");
```

SecondActivity 对应的代码如下。

```
public class SecondActivity extends Activity {
    private Button button;
    @Override
    protected void onCreate(Bundle savedInstanceState){
        //TODO Auto-generated method stub
        super.onCreate(savedInstanceState);
        setContentView(R.layout.second_layout);
        System.out.println("SecondActivity:onCreate");
        button= (Button)this.findViewById(R.id.button1);
        button.setOnClickListener(new OnClickListener(){
            @Override
            public void onClick(View arg0){
```

```java
            finish();                    //关闭 Activity        }
    });
}

@Override
protected void onDestroy(){
    //TODO Auto-generated method stub
    super.onDestroy();
    System.out.println("SecondActivity:onDestroy");
}

@Override
protected void onPause(){
    //TODO Auto-generated method stub
    super.onPause();
    System.out.println("SecondActivity:onPause");
}

@Override
protected void onRestart(){
    //TODO Auto-generated method stub
    super.onRestart();
    System.out.println("SecondActivity:onReStart");
}

@Override
protected void onResume(){
    //TODO Auto-generated method stub
    super.onResume();
    System.out.println("SecondActivity:onResume");
}

@Override
protected void onStart(){
    //TODO Auto-generated method stub
    super.onStart();
    System.out.println("SecondActivity:onStart");
}

@Override
protected void onStop(){
    //TODO Auto-generated method stub
    super.onStop();
    System.out.println("SecondActivity:onStop");
```

}

}

同样地，在 MainActivity 中，重写 SecondActivity 类中的 Activity 的几个生命周期方法，在每个方法的方法体里输出当前所在 Activity 名：当前所在方法名。MainActivity 对应的代码如下。

```java
import android.app.Activity;
import android.content.Intent;
import android.os.Bundle;
import android.view.View;
import android.view.View.OnClickListener;
import android.view.Window;
import android.view.WindowManager;
import android.widget.Button;

public class MainActivity extends Activity {
    private Button button;

    @Override
    protected void onCreate(Bundle savedInstanceState){
        super.onCreate(savedInstanceState);
        requestWindowFeature(Window.FEATURE_NO_TITLE);
        setContentView(R.layout.activity_main);
        getWindow(). setFlags ( WindowManager. LayoutParams. FLAG _ FULLSCREEN,
        WindowManager.LayoutParams.FLAG_FULLSCREEN);
        System.out.println("MainActivity:onCreate");
        button=(Button)this.findViewById(R.id.button1);
        button.setOnClickListener(new OnClickListener(){
            @Override
            public void onClick(View arg0){
                //生成一个意图对象
                Intent intent=new Intent();
                //设置意图
                intent.setClass(MainActivity.this,SecondActivity.class);
                //启动 Activity
                startActivity(intent);
            }
        });
    }
    @Override
    protected void onStart(){
```

```java
        //TODO Auto-generated method stub
        super.onStart();
        System.out.println("MainActivity:onStart");
    }

     @Override
    protected void onRestart(){
        //TODO Auto-generated method stub
        super.onRestart();
        System.out.println("MainActivity:onRestart");
    }
     @Override
    protected void onResume(){
        //TODO Auto-generated method stub
        super.onResume();
        System.out.println("MainActivity:onResume");
    }
     @Override
    protected void onPause(){
        //TODO Auto-generated method stub
        super.onPause();
        System.out.println("MainActivity:onPause");
    }
     @Override
    protected void onStop(){
        //TODO Auto-generated method stub
        super.onStop();
        System.out.println("MainActivity:onStop");
    }
     @Override
    protected void onDestroy(){
        //TODO Auto-generated method stub
        super.onDestroy();
        System.out.println("MainActivity:onDestroy");
    }

}
```

MainActivity 对应的布局文件 activity_main 代码如下。

```xml
<RelativeLayout xmlns:android="http://schemas.android.com/apk/res/android"
    xmlns:tools="http://schemas.android.com/tools"
    android:layout_width="match_parent"
    android:layout_height="wrap_content"
    tools:context=".MainActivity">
```

```xml
<TextView
        android:id="@+id/textView1"
        android:layout_width="wrap_content"
        android:layout_height="wrap_content"
        android:text="第一个 Activity"/>

<Button
        android:id="@+id/button1"
        android:layout_width="wrap_content"
        android:layout_height="wrap_content"
        android:layout_alignParentLeft="true"
        android:layout_below="@+id/textView1"
        android:layout_marginTop="18dp"
        android:text="跳到第二个 Activity"/>
</RelativeLayout>
```

第一个 Activity 的运行效果如图 3.11 所示。

图 3.11　第一个 Activity 运行效果

第二个 Activity 的运行效果如图 3.12 所示。

图 3.12　第二个 Activity 运行效果

使用 Logcat 观察输出结果，如图 3.13 所示。

cn.edu.neusoft.software.mobile.activitydemo I/System.out: MainActivity:onCreate
cn.edu.neusoft.software.mobile.activitydemo I/System.out: MainActivity:onStart
cn.edu.neusoft.software.mobile.activitydemo I/System.out: MainActivity:onResume
cn.edu.neusoft.software.mobile.activitydemo I/System.out: MainActivity:onPause
cn.edu.neusoft.software.mobile.activitydemo I/System.out: SecondActivity:onCreate
cn.edu.neusoft.software.mobile.activitydemo I/System.out: SecondActivity:onStart
cn.edu.neusoft.software.mobile.activitydemo I/System.out: SecondActivity:onResume
cn.edu.neusoft.software.mobile.activitydemo I/System.out: MainActivity:onStop
cn.edu.neusoft.software.mobile.activitydemo I/System.out: SecondActivity:onPause
cn.edu.neusoft.software.mobile.activitydemo I/System.out: MainActivity:onRestart
cn.edu.neusoft.software.mobile.activitydemo I/System.out: MainActivity:onStart
cn.edu.neusoft.software.mobile.activitydemo I/System.out: MainActivity:onResume
cn.edu.neusoft.software.mobile.activitydemo I/System.out: SecondActivity:onStop

图 3.13　Logcat 输出结果

3.4.3 初识 Intent：在不同 Activity 之间传递数据

在一个 Activity 中启动另一个 Activity 时，经常需要传递一些数据。这时就可以通过 Intent 来实现，因为 Intent 通常被称为两个 Activity 之间的信使。将要传递的数据保存在 Intent 中，就可以将其传递到另一个 Activity 中了。

在 Android 中，通过 Intent 提供的 putExtras()方法，可以将要携带的数据保存到 Intent 中。然后再通过 getExtras()方法获取 Intent 中存放的数据。下面通过一个具体实例介绍使用 Intent 在 Activity 之间交换数据的方法。

【例 3-3】 使用 Intent 从 TransmitDataActivity 传递一个字符串、一个整数和一个 Data 对象到 MyActivity 中，在 MyActivity 中把传递过来的数据显示在 TextView 上。

将数据保存到 Intent 对象，并跳到另一个用来显示这些数据的 Activity 的代码如下。

```
import android.content.Intent;
import android.os.Bundle;
import android.support.v7.app.AppCompatActivity;
import android.view.View;
import android.widget.Button;
import cn.edu.neusoft.software.mobile.activitydemo.R;
public class TransmitDataActivity extends AppCompatActivity {
    private Button btn;
    @Override
    protected void onCreate(Bundle savedInstanceState){
        super.onCreate(savedInstanceState);
        setContentView(R.layout.activity_transmit_data);
        btn= (Button)findViewById(R.id.btnValue);
        btn.setOnClickListener(new View.OnClickListener(){
            @Override
            public void onClick(View v){
                Intent intent=new Intent(TransmitDataActivity.this,MyActivity.class);
                intent.putExtra("intent_string","使用 intent 传值");
                intent.putExtra("intent_integer",100);
                Data data=new Data();
                data.id=1000;
                data.name="Android";
                intent.putExtra("intent_object",data);
                startActivity(intent);
            }
        });
    }
}
```

以上代码涉及一个 Data 类，这个类是可序列化的，也就是实现了 java.io.Serializable

接口的类。Data 类的代码如下。

```java
public class Data implements Serializable{
    String name;
    int id;
}
```

在 MyActivity 类中获取通过 Intent 对象传递来的 3 个值（String、Integer 和 Data 类型的值）的代码如下。

```java
import android.os.Bundle;
import android.support.v7.app.AppCompatActivity;
import android.widget.TextView;
import cn.edu.neusoft.software.mobile.activitydemo.R;
public class MyActivity extends AppCompatActivity {

    @Override
    protected void onCreate(Bundle savedInstanceState){
        super.onCreate(savedInstanceState);
        setContentView(R.layout.activity_my);
        TextView textView= (TextView)findViewById(R.id.textResult);

        String intentString=getIntent().getStringExtra("intent_string");
        /* String intentString = getIntent().getExtras().getString("intent_string");*/
        int intentInteger=getIntent().getExtras().getInt("intent_integer");
        Data data= (Data)getIntent().getExtras().get("intent_object");

        StringBuffer sb=new StringBuffer();
        sb.append("intent_string");
        sb.append(intentString);
        sb.append("\n");
        sb.append("intent_integer");
        sb.append(intentInteger);
        sb.append("\n");
        sb.append("data.id");
        sb.append(data.id);
        sb.append("\n");
        sb.append("data.name");
        sb.append(data.name);
        sb.append("\n");
        textView.setText(sb.toString());
    }
}
```

程序分析：使用 Intent 传递数据，是最常用的一种数据传递的方法。通过 Intent.

putExtra()方法可以将简单类型的数据或可序列化的对象保存在 Intent 对象中,然后在目标 Activity 中使用 getXxx(getInt、getString 等)方法获得这些数据。

运行程序后,单击图 3.14 所示的按钮,会显示图 3.15 所示的输出信息。

图 3.14　TransmitData 程序的主界面　　　图 3.15　显示从 Intent 对象中获取的数据

3.5　知识拓展:Activity/ActionBarActivity/AppCompatActivity

在 Android Studio 中创建的 MainActivity 类,默认是继承自 AppCompatActivity,它是 Activity 的子类。ActionBarActivity 与 AppCompatActivity 都是 Activity 的子类,ActionBarActivity 在 Android Studio 中已经是过期类。如果在 Android Studio 中创建的类继承自 Activity,则不带 ActionBar。如果要在 Android Studio 中也使用 ActionBar,并且不使用已经过时的 ActionBarActivity,有什么办法呢? 就是使用 AppCompatActivity。在 Android Studio 中,把 MainActivity 继承自 AppCompatActivity,并导入对应的包。以后在项目中,可以通过手动修改 Activity 的继承父类来决定是否显示 ActionBar,并且对程序没有其他影响。

本章小结

本章给出了一个简单例子,作为 Android 入门学习的第一个程序。通过这个程序,读者可以掌握建立 Android 程序的方法,理解 Android 程序的基本结构,理解 Android 程序的组件式开发方法,掌握 Android 的四大组件及 Activity 的基本概念与用法。

本章练习

1. 两个 Activity 之间跳转时,必然会执行的是哪几个方法?
2. 简述 Android 系统的 4 种基本组件 Activity、Service、BroadcastReceiver 和 ContentProvider 的用途。
3. Intent 的作用是什么? 传递数据时,Intent 可以传递哪些类型数据?
4. 简述 AndroidManifest.xml 文件的用途。
5. 简要说明一个 Android 工程的结构及 AndroidManifest.xml 的作用。
6. 简要说明 gradle 文件的主要作用。
7. Activity 对象的 7 个生命周期方法是什么? 说明各个方法的执行时机。
8. Activity 生命周期中,第一个需要执行的方法是什么?

第 4 章

CoffeeStore 项目导学

本章概述

通过本章的学习,读者要整体了解项目的基本功能,了解项目的体系结构设计、原型设计以及数据库设计。

学习重点与难点

重点:

(1) CoffeeStore 的基本功能。
(2) CoffeeStore 的体系结构。
(3) CoffeeStore 的原型设计。
(4) CoffeeStore 的数据库设计。

难点:

(1) CoffeeStore 的体系结构。
(2) CoffeeStore 的原型设计。

学习建议

读者可以了解 CoffeeStore 项目的基本功能及界面原型,后续学到某个功能实现方法时可以再回头阅读本章内容,加深理解。

4.1 功能描述

商城类 App 一般分为商品浏览、商品详情查看、加入购物车、商品购买、生成订单、订单查看等流程。在线咖啡销售系统的用户有顾客和管理员两种角色。管理员的主要功能有用户信息管理、咖啡管理、咖啡种类管理、评论管理、商品搜索、销售统计等。管理员的功能主要通过 Web 后台管理系统来实现。顾客端功能主要在手机端实现。用户可以通过手机 App 实现个人账号管理、登录、商品分类信息查看、商品详细信息查看、推荐商品、商品搜索、个人收藏、个人订单管理、发表评论、商品结算等功能,完成个人日常生活中通过手机购买咖啡的需求。CoffeeStore 的功能结构如图 4.1 所示,图中虚线内部为手机顾客端主要实现的功能。

第 4 章 CoffeeStore 项目导学

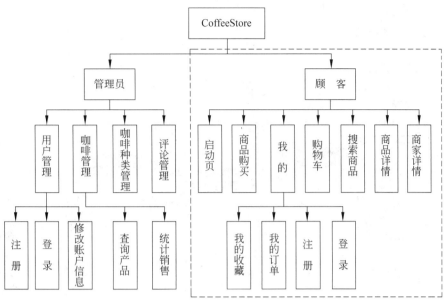

图 4.1　功能结构图

4.2　体系结构与知识点

本项目是典型的移动互联网应用开发项目,这种项目的基本程序框架如图 4.2 所示。移动互联网是以移动网络作为接入网络的互联网及其服务,包括三个要素:移动终端、移动网络和应用服务,其中应用服务是移动互联网的核心,也是用户的最终目的。

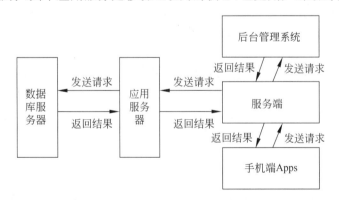

图 4.2　移动互联网应用程序体系结构

CoffeeStore 是一种典型的移动互联网应用系统,本系统分为数据库服务器、应用服务器、服务器端和移动手机端,其系统整体结构如图 4.2 所示。当手机端需要显示各类信息或有数据需要更新时,手机端向服务器发送请求,服务器端接到请求后处理请求,然后把结果返回给手机客户端。

CoffeeStore 项目基本涵盖了 Android 应用开发的重要知识点,完成这个项目涉及的

Android 开发主要技术点如图 4.3 所示。

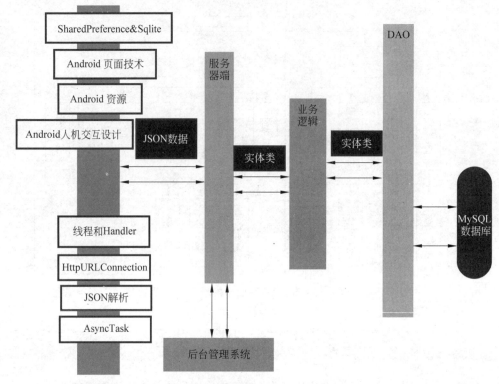

图 4.3　本项目涉及的知识点

由图 4.3 可以看出，系统大量的商品信息和用户信息存储到远程服务器上，远程服务器采用 MySQL 数据库。由于手机的存储容量有限，所有手机上的信息都来自服务器，每当手机端需要从数据库提取信息或更新数据库信息时，首先要向服务器发送请求，服务器收到请求后，到后台数据库提取或更新相应信息，然后把操作结果返回给手机客户端。服务端与客户端采用 Socket 方式进行通信，数据格式采用 JSON。手机端仅存储少量信息，个人用户的基本信息可采用 SharedPreference 存储，个人订单及购物车可采用本地数据库 SQLite 存储。

涉及 Android 客户端的技术如下。

① 页面技术：布局管理器与基本控件、ListView、Gridview、Gallery、ScrollView 等高级控件、自定义适配器、自定义控件、引用第三方控件、Fragment 与 ActionBar、字符串资源、主题与风格、Android 动画、尺寸资源、颜色资源、数组资源、菜单与对话框、WebView 控件、本地化与国际化等。

② 与服务器通信的数据格式为 JSON。

③ 网络通信技术：HttpURLConnection、集合、线程与 Handler、异步任务类、JSON 数据解析、XML 解析、WebService。

④ 本地存储技术：SDCard、SharedPreference 与 SQLite。

4.3 原型设计

本项目的界面原型设计如图 4.4~图 4.9 所示。图 4.4 为首页,首页分为广告轮播区、分类列表区、推荐商品区、热门商品区及打折商品区。图 4.5 为店铺列表页面,列出了一些店铺信息,图 4.6 为咖啡商品页面,包括商品的图片、商品名称、商品简介、商品价格以及商品的星级评分。图 4.7 为登录页面,若想购买商品及查看订单,首先需要登录。图 4.8 为"我的"页面,登录后可查看我的基本信息。图 4.9 为意见反馈页面,可以反馈对软件的使用意见信息。

图 4.4 首页

图 4.5 店铺列表页

图 4.6 咖啡列表页

图 4.7 登录页面

图 4.8 "我的"页面

图 4.9 意见反馈页面

4.4 数据库设计

本项目的大量数据存储在远程数据库中,远程数据库采用 MySQL 数据库,"我的收藏"数据来源于本地,本地数据库采用 SQLite 数据库。

本项目的远程数据库有 6 张表,本地数据库有 1 张表,存放用户收藏的店铺列表存在本地。远程数据库的表设计如表 4.1～表 4.6 所示,实现远程服务功能时可参考。

表 4.1 用户信息表(t_users)

字 段 描 述	数 据 模 型	主 键	可否为空	描 述
user_id(用户编号)	int(11)	是		
user_name(用户名)	varchar(50)			
user_password(用户密码)	varchar(50)			
user_sex(用户性别)	TINYINT			
user_birthday(用户生日)	char(20)			
user_phone(用户电话)	char(11)			

表 4.2 用户收货地址信息表(t_address)

字 段 描 述	数 据 模 型	主 键	可否为空	描 述
address_id(地址编号)	int(11)	是		
address_name(收货人姓名)	varchar(20)			

续表

字段描述	数据模型	主键	可否为空	描述
address_phone(收货人电话)	char(11)			
address_postal(收货地址邮编)	char(6)			
address_region(收货区域)	varchar(50)			
address_detail(收货详细地址)	varchar(50)			
user_id(用户编号)	int(11)			

表 4.3 咖啡表（t_coffee）

字段描述	数据模型	主键	可否为空	描述
coffee_id(咖啡编号)	int(11)	是		
coffee_name(咖啡名称)	varchar(50)			
coffee_price(咖啡价格)	float(10)			
coffee_intro(咖啡简介)	varchar(255)			
coffee_com(咖啡评论)	TEXT			
coffee_image(咖啡图片)	BLOB			
style_id(类别编号)	int(11)			

表 4.4 订单表（t_orders）

字段描述	数据类型	主键	可否为空	描述
order_id(商品标号)	int(11)	是		
coffee_count(咖啡数量)	int(11)			
pay_date(收货日期)	DATE			
pay_time(时间)	DATE			
pay_way(方式)	varchar(20)			
user_id(用户编号)	int(11)			
coffee_id(咖啡编号)	int(11)			
address_id(地址编号)	int(11)			
is_paid(是否付款)	TINYINT			

表 4.5 反馈表（t_feedback）

字段描述	数据类型	主键	可否为空	描述
feedback_id(反馈编号)	int(11)	是		
feedback_type(反馈类型)	varchar(10)			
feedback_content(反馈内容)	TEXT			
feedback_commu(反馈存根)	TEXT			

表 4.6 咖啡类别表（t_style）

字 段 描 述	数据类型	主 键	可否为空	描 述
style_id（类别编号）	int(11)	是		
style_name（类别名称）	varchar(50)			

本地数据库存放收藏夹里的店铺列表，表的结构如表 4.7 所示。

表 4.7 店铺收藏夹表（t_favoriate）

字 段 描 述	数据类型	主 键	可否为空	描 述
shop_id（类别编号）	int(11)	是		
shop_name（类别名称）	varchar(50)			
shop_adress（店铺地址）				
shop_adress（店铺电话）				
shop_adress（店铺图片）				

本 章 小 结

通过本章学习，读者可以清晰地了解 CoffeeStore 的基本功能、项目构思和设计情况以及涉及的 Android 知识点，从而对项目的整体情况有清晰的认识，明确后续实施和运行中的任务。

本 章 习 题

1. CoffeeStore 的基本功能有哪些？
2. 远程数据库存储用了几张表？表的基本结构是什么样的？
3. 移动互联网程序的基本体系结构是什么？
4. CoffeeStore 项目中本地数据库存储的是什么信息？

第 2 篇

界面开发——Android 界面开发篇

项目导引

一个 App 的界面设计至关重要。界面是人机交互窗口,是一个软件的门户,应该兼具美观、实用、友好等特点。本书中的界面设计也是十分考究的,汇集了各种功能的首页是 App 界面设计中最复杂的一个。首页是一个手机 App 界面的核心所在,包含了很多功能键,有菜单栏、状态栏、分类等一系列功能,因此它的设计至关重要。本项目中的首页,有的是图文并茂,有的是直接用图片拼接,可以说是技术和美术的结合。多个页面都起到至关重要的作用,兼具美观性和功能性特点,这些页面是如何构建起来的,其中页面的组成部分——组件又是如何友好、有序、美观地排列于界面之上的呢?这正是本篇章要介绍的主要内容。

首先来看看 CoffeeStore 首页的界面设计,图Ⅱ.1 是首页界面的效果图。

图中应用了多种布局,包括线性布局、网格布局、绝对布局等,也应用了多种界面组件,包括 TextView、Button、ScrollView 等。

图Ⅱ.2 是店铺列表界面,这个店铺列表界面也使用了多种布局和组件。

界面布局(Layout)是用户界面结构的描述,定义了界面中所有的元素、结构和相互关系。

本篇将通过实现 CoffeeStore 的各个界面讲解 Android 的用户界面构建。Android 用户界面构建包括如下内容。

- Android 布局管理器

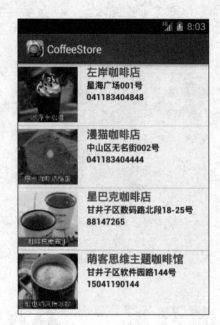

图Ⅱ.1 首页界面图　　　　　图Ⅱ.2 店铺列表界面

- Android 基本控件
- ViewPager 与 Fragment
- Android 高级控件
- 资源样式与主题
- Android 人机交互设计

Android 布局管理器

本章概述

通过本章的学习，读者应掌握 Android 的布局管理器，实现界面组件的合理布局，包括线性布局管理器、相对布局管理器、表格布局管理器、网格布局管理器、帧布局管理器、布局管理器的嵌套与综合运用，向容器中手动添加控件。

学习重点与难点

重点：

（1）线性布局管理器。

（2）相对布局管理器。

（3）表格布局管理器。

（4）网格布局管理器。

（5）帧布局管理器。

难点：

（1）布局管理器的嵌套与综合运用。

（2）向容器中手动添加控件。

学习建议

读者在学习中要深入理解 Android 中各个布局管理器的常用参数，多看案例，多思考，多动手实践，熟练地综合应用各个布局管理器，从而搭建出需要实现的界面。

界面布局（Layout）是用户界面结构的描述，定义了界面中所有的元素、结构和相互关系。在 Android 中，每个控件在窗体中都有具体的位置和尺寸，在窗体中摆放各种控件时，很难准确判断控件的具体位置和大小，而使用 Android 的布局管理器可以很方便地控制各个控件的位置和大小。Android 中的布局是一个容器，在此容器中可放置其他控件，大部分容器控件继承于 ViewGroup 类，Android 中提供了线性布局、相对布局、表格布局、网格布局、帧布局五种布局管理器。对应这五种布局管理器，Android 提供了五种布局方式，下面就依次讲解这五种布局，为完成 CoffeeStore 项目打下基础。

5.1 线性布局管理器

线性布局(LinearLayout)是最简单的一种布局,这种布局比较常用,它按照垂直或者水平的顺序依次排列子元素,每一个子元素都位于前一个元素之后。如果是垂直排列,那么将是一个 N 行单列的结构,每一行只会有一个元素,不限制元素的宽度;如果是水平排列,那么将是一个单行 N 列的结构。如果搭建两行两列的结构,通常的方式是先垂直排列两个元素,每一个元素里再包含一个 LinearLayout,进行水平排列。

线性布局有 4 个重要的参数,决定元素的布局和位置,这 4 个参数如下。

① android:layout_weight:线性布局内子元素对未占用空间(水平或垂直)分配权重值,值越小,权重越大。

② android:orientation:线性布局以列或行来显示内部子元素。

③ android:layout_gravity:是本元素相对于父元素的重力方向。

④ android:gravity:是本元素所有子元素的重力方向。

首先介绍 LinearLayout 中的子元素属性 android:layout_weight,它用于描述该子元素在剩余空间中占有的大小比例。假如一行只有一个文本框,那么它的默认值就为 0,如果一行中有两个等长的文本框,那么它们的 android:layout_weight 值可以同为 1。如果一行中有两个不等长的文本框,假设第一个文本框与第二个文本框的 android:layout_weight 值分别为 1 和 2,在文本框宽度设为 match_parent 的情况下,第一个文本框将占据剩余空间的 2/3,第二个文本框将占据剩余空间的 1/3。在文本框的宽度设为 0dp 或者 wrap_content 的情况下,则第一个文本框将占据剩余空间的 1/3,第二个文本框将占据剩余空间的 2/3。

为单独的子元素指定 weight 值,其好处就是允许子元素填充屏幕上的剩余空间。weight 的默认值为 0,为子元素指定一个 weight 值,剩余的空间就会按这些子元素指定的 weight 比例分配给这些子元素。

LinearLayout 属性中的 android:orientation 属性是设置线性布局方式的。LinearLayout 线性布局分为水平线性布局和垂直线性布局,当其值为 vertical 时,为垂直线性布局,当其值为 horizontal 时,为水平线性布局。不管是水平还是垂直线性布局,一行(列)只能放置一个控件。通过下面的代码来看一下 android:orientation 属性的应用。

```
<LinearLayout xmlns:android="http://schemas.android.com/apk/res/android"
android:layout_width="match_parent"
android:layout_height="match_parent"
android:orientation="vertical" >
<TextView
    android:layout_width="match_parent"
    android:layout_height="wrap_content"
    android:text="Link to Bluetooth?" />
</LinearLayout>
```

以上代码中使用线性布局,其中 android:orientation 用来表明该布局中控件的显示方向。vertical 表示控件为垂直方向显示。如果想显示为水平方向,只需要将属性值改为 horizontal。

最后介绍 android:layout_gravity 和 android:gravity 的使用。android:gravity 是对元素本身说的,元素本身的文本靠左还是靠右显示主要由 gravity 属性设置,不设置默认是在左侧。android:layout_gravity 是相对于它的父元素说的,说明元素显示在父元素的什么位置。比如,button:android:layout_gravity 表示按钮在界面上的位置。android:gravity 表示 button 上的字在 button 上的位置。

这两个属性的可选值也是相同的,包括 top、bottom、left、right、center_vertical、fill_vertical、center_horizontal、fill_horizontal、center、fill、clip_vertical、clip_horizoutal。这些属性是可以多选的,用 | 分开。默认的值是 Gravity.LEFT。对这些属性的描述如表 5.1 所示。

表 5.1　android:layout_gravity 和 android:gravity 属性取值

取　　值	说　　明
top	将对象放在其容器的顶部,不改变其大小
bottom	将对象放在其容器的底部,不改变其大小
left	将对象放在其容器的左侧,不改变其大小
right	将对象放在其容器的右侧,不改变其大小
center_vertical	将对象纵向居中,不改变其大小;垂直对齐方式:垂直方向上居中对齐
fill_vertical	必要时增加对象的纵向大小,以完全充满其容器;垂直方向填充
center_horizontal	将对象横向居中,不改变其大小;水平对齐方式:水平方向上居中对齐
fill_horizontal	必要时增加对象的横向大小,以完全充满其容器;水平方向填充
center	将对象横纵居中,不改变其大小
fill	必要时增加对象的横纵向大小,以完全充满其容器
clip_vertical	附加选项,用于按照容器的边来剪切对象的顶部和/或底部的内容。剪切基于其纵向对齐设置:顶部对齐时剪切底部;底部对齐时剪切顶部;除此之外剪切顶部和底部;垂直方向裁剪
clip_horizontal	附加选项,用于按照容器的边来剪切对象的左侧和/或右侧的内容。剪切基于其横向对齐设置:左侧对齐时剪切右侧;右侧对齐时剪切左侧;除此之外剪切左侧和右侧;水平方向裁剪

声明 Android 程序的界面布局有两种方法。

(1) 使用 XML 文件描述界面布局(推荐使用)。

(2) 在程序运行时动态添加或修改界面布局。

既可以独立使用任何一种声明界面布局的方式,也可以同时使用两种方式。

【例 5-1】　线性布局管理器。

实现步骤如下。

新建一个模块,名字为 ch05_1,打开项目文件夹中 res\layout 目录下的 activity_main.xml 文件,界面代码如下。

```xml
<?xml version="1.0" encoding="utf-8"?>
<LinearLayout xmlns:android="http://schemas.android.com/apk/res/android"
    android:layout_width="match_parent"
    android:layout_height="match_parent"
    android:orientation="vertical">

    <LinearLayout
        android:layout_width="match_parent"
        android:layout_height="match_parent"
        android:layout_weight="1"
        android:orientation="vertical">

        <LinearLayout
            android:layout_width="match_parent"
            android:layout_height="match_parent"
            android:layout_weight="1"
            android:orientation="horizontal">

            <LinearLayout
                android:layout_width="match_parent"
                android:layout_height="match_parent"
                android:layout_weight="1"
                android:orientation="vertical">

                <Button
                    android:layout_width="wrap_content"
                    android:layout_height="wrap_content"
                    android:layout_gravity="left"
                    android:text="左上按钮"/>
            </LinearLayout>

            <LinearLayout
                android:layout_width="match_parent"
                android:layout_height="match_parent"
                android:layout_weight="1"
                android:orientation="vertical">

                <Button
                    android:layout_width="wrap_content"
                    android:layout_height="wrap_content"
                    android:layout_gravity="right"
```

```xml
            android:text="右上按钮"/>
    </LinearLayout>
</LinearLayout>

<LinearLayout
    android:layout_width="match_parent"
    android:layout_height="match_parent"
    android:layout_weight="1"
    android:gravity="center"
    android:orientation="vertical">

    <Button
        android:layout_width="wrap_content"
        android:layout_height="wrap_content"
        android:text="中心按钮"/>
</LinearLayout>

<LinearLayout
    android:layout_width="match_parent"
    android:layout_height="match_parent"
    android:layout_weight="1"
    android:orientation="horizontal">

    <LinearLayout
        android:layout_width="match_parent"
        android:layout_height="match_parent"
        android:layout_weight="1"
        android:gravity="left|bottom"
        android:orientation="vertical">

        <Button
            android:layout_width="wrap_content"
            android:layout_height="wrap_content"
            android:text="左下按钮"/>
    </LinearLayout>

    <LinearLayout
        android:layout_width="match_parent"
        android:layout_height="match_parent"
        android:layout_weight="1"
        android:gravity="right|bottom"
        android:orientation="vertical">

        <Button
```

```xml
            android:layout_width="wrap_content"
            android:layout_height="wrap_content"
            android:text="右下按钮"/>
        </LinearLayout>
    </LinearLayout>
</LinearLayout>

<LinearLayout
    android:layout_width="match_parent"
    android:layout_height="match_parent"
    android:layout_weight="1"
    android:orientation="vertical">

    <ImageView
        android:layout_width="match_parent"
        android:layout_height="match_parent"
        android:layout_weight="1"
        android:src="@drawable/background"/>

    <EditText
        android:layout_width="match_parent"
        android:layout_height="wrap_content"
        android:hint="请在这里输入文本"/>
    </LinearLayout>
</LinearLayout>
```

以上代码中应用了多个 LinearLayout 布局,具体包含关系如图 5.1 所示。

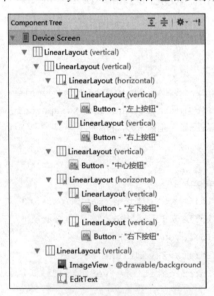

图 5.1 线性布局管理器布局结构图

运行程序,效果如图 5.2 所示。

图 5.2　线性布局管理器运行效果图

5.2　相对布局管理器

相对布局由 RelativeLayout 表示。相对布局容器内子组件的位置总是由相对兄弟组件、父容器的位置来决定的,因此这种布局方式称为相对布局。如果 A 组件的位置是由 B 组件的位置来决定的,Android 要求先定义 B 组件,再定义 A 组件。

RelativeLayout 按照各子元素之间的位置关系完成布局。在此布局中,子元素中与位置相关的属性将生效。具体的属性如下。

第一类:属性值为 true 或 false。

android:layout_centerHrizontal　水平居中

android:layout_centerVertical　垂直居中

android:layout_centerInparent　相对于父元素完全居中

android:layout_alignParentBottom　贴紧父元素的下边缘

android:layout_alignParentLeft　贴紧父元素的左边缘

android:layout_alignParentRight　贴紧父元素的右边缘

android:layout_alignParentTop　贴紧父元素的上边缘

android:layout_alignWithParentIfMissing　如果对应的兄弟元素找不到,就以父元素

作参照物

第二类：属性值必须为ID的引用名"@id/id-name"。

android:layout_below 在某元素的下方

android:layout_above 在某元素的上方

android:layout_toLeftOf 在某元素的左边

android:layout_toRightOf 在某元素的右边

android:layout_alignTop 本元素的上边缘和某元素的上边缘对齐

android:layout_alignLeft 本元素的左边缘和某元素的左边缘对齐

android:layout_alignBottom 本元素的下边缘和某元素的下边缘对齐

android:layout_alignRight 本元素的右边缘和某元素的右边缘对齐

第三类：属性值为具体的像素值，如40dip、30px。

android:layout_marginBottom 离某元素底边缘的距离

android:layout_marginLeft 离某元素左边缘的距离

android:layout_marginRight 离某元素右边缘的距离

android:layout_marginTop 离某元素上边缘的距离

android:gravity 属性是用来设置元素的文本内容的显示位置，比如一个按钮上面的文本可以靠左或者靠右显示。如果要设置按钮的文本靠右显示，就编写为android:gravity="right"。

android:layout_gravity 是用来设置该元素相对于其父类元素的位置，比如一个按钮显示在线性布局管理器中。如果要设置按钮靠右显示，就编写为android:layout_gravity="right"。

注意，在指定位置关系时，引用的ID必须在引用之前先被定义，否则将出现异常。

相对布局是Android五大布局结构中最灵活的一种布局结构，比较适合一些复杂界面的布局。

相对布局的语法结构如下。

```
<RelativeLayout
    android:layout_width="match_parent"
    android:layout_height="220dp"
    android:background="#F00"
    android:orientation="vertical">
<!--相对父容器 -->
<!-添加需要的控件-->
</RelativeLayout>
```

下面通过一个实例来学习相对布局的应用。

【例5-2】 相对布局管理器。

实现步骤如下。

新建一个模块，名字为ch05_2，打开项目文件夹中res\layout目录下的activity_main.xml文件，将其中的代码替换为以下代码。

```xml
<?xml version="1.0" encoding="utf-8"?>
<LinearLayout xmlns:android="http://schemas.android.com/apk/res/android"
    android:layout_width="match_parent"
    android:layout_height="match_parent"
    android:orientation="vertical" >

<RelativeLayout
        android:layout_width="match_parent"
        android:layout_height="220dp"
        android:background="#f40808"
        android:orientation="vertical">
<!--相对父容器 -->

<Button
          android:layout_width="wrap_content"
          android:layout_height="wrap_content"
          android:text="B1"
          android:layout_alignParentLeft="true"
          android:layout_alignParentTop="true" />
<Button
          android:layout_width="wrap_content"
          android:layout_height="wrap_content"
          android:text="B2"
          android:layout_alignParentLeft="true"
          android:layout_alignParentBottom="true" />
<Button
          android:layout_width="wrap_content"
          android:layout_height="wrap_content"
          android:text="B3"
          android:layout_alignParentRight="true"
          android:layout_alignParentTop="true" />
<Button
          android:layout_width="wrap_content"
          android:layout_height="wrap_content"
          android:text="B4"
          android:layout_alignParentRight="true"
          android:layout_alignParentBottom="true" />
<Button
          android:layout_width="wrap_content"
          android:layout_height="wrap_content"
          android:text="B5"
          android:layout_alignParentLeft="true"
```

```xml
                android:layout_centerVertical="true"/>
    <Button
                android:layout_width="wrap_content"
                android:layout_height="wrap_content"
                android:text="B6"
                android:layout_alignParentRight="true"
                android:layout_centerVertical="true"/>

    <Button
                android:layout_width="wrap_content"
                android:layout_height="wrap_content"
                android:text="B7"
                android:layout_alignParentTop="true"
                android:layout_centerHorizontal="true"/>
    <Button
                android:layout_width="wrap_content"
                android:layout_height="wrap_content"
                android:text="B8"
                android:layout_alignParentBottom="true"
                android:layout_centerHorizontal="true"/>

    <Button
                android:layout_width="wrap_content"
                android:layout_height="wrap_content"
                android:text="B9"
                android:layout_centerInParent="true" />
</RelativeLayout>
<RelativeLayout
        android:layout_width="match_parent"
        android:layout_height="220dp"
        android:background="#1ff017"
        android:orientation="vertical">

    <TextView
                android:layout_width="wrap_content"
                android:layout_height="wrap_content"
                android:textSize="24sp"
                android:text="登录"
                android:layout_alignParentTop="true"
                android:layout_centerHorizontal="true"
                android:id="@+id/lblTitle"/>

    <EditText
```

```
            android:layout_width="250dp"
            android:layout_height="wrap_content"
            android:layout_alignParentRight="true"
            android:layout_below="@id/lblTitle"
            android:id="@+id/txtUserName"/>

<EditText
            android:layout_width="250dp"
            android:layout_height="wrap_content"
            android:layout_alignParentRight="true"
            android:layout_below="@id/txtUserName"
            android:id="@+id/txtPassword"/>

<Button
            android:layout_width="120dp"
            android:layout_height="wrap_content"
            android:layout_alignParentRight="true"
            android:layout_below="@id/txtPassword"
            android:text="Cancel"
            android:id="@+id/btnCancel" />
<Button
            android:layout_width="120dp"
            android:layout_height="wrap_content"
            android:layout_toLeftOf="@id/btnCancel"
            android:layout_below="@id/txtPassword"
            android:text="OK"
            android:id="@+id/btnOK" />
<TextView
            android:layout_width="70dp"
            android:layout_height="wrap_content"
            android:gravity="right"
            android:layout_alignBaseline="@id/txtUserName"
            android:layout_alignParentLeft="true"
            android:text="UserName"
            android:id="@+id/lblUserName"/>
<TextView
            android:layout_width="70dp"
            android:layout_height="wrap_content"
            android:gravity="right"
            android:layout_alignBaseline="@id/txtPassword"
            android:layout_alignParentLeft="true"
            android:text="Password"
            android:id="@+id/lblPassword"/>"
```

```
</RelativeLayout>
</LinearLayout>
```

程序的布局结构如图 5.3 所示。

由图 5.3 可见,本案例中用到两次相对布局管理器,两个相对布局并列排放于一个线性布局内。第一个相对布局包括一组数量为 9 个的按钮,另外一组包括 7 个控件,是一个典型的登录页面的组成。

程序运行效果如图 5.4 所示。

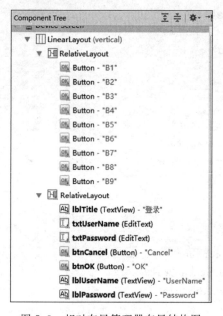

图 5.3　相对布局管理器布局结构图

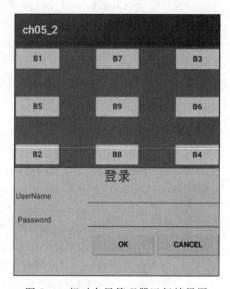

图 5.4　相对布局管理器运行效果图

5.3　表格布局管理器

表格布局(TableLayout)是一种常用的界面布局,通过指定行和列将界面元素添加到表格中,表格的边界对用户是不可见的。表格布局支持嵌套,就是可以将表格布局放置在表格布局的表格中,也可以在表格布局中添加其他界面布局,如线性布局、相对布局等,而且表格布局可以跨行,但是不能跨列。

表格布局通常适用于 N 行 N 列的布局格式。一个 TableLayout 由许多 TableRow 组成,一个 TableRow 就代表 TableLayout 中的一行。TableRow 是 LinearLayout 的子类,TableLayout 并不需要明确地声明包含多少行、多少列,而是通过 TableRow 以及其他组件来控制表格的行数和列数。TableRow 也是容器,因此可以向 TableRow 里面添加其他组件,每添加一个组件,该表格就增加一列。如果想在 TableLayout 里面添加组件,该组件就直接占用一行。在表格布局中,列的宽度由该列中最宽的单元格决定,整个表格布局的宽度取决于父容器的宽度(默认是占满父容器本身)。

TableLayout 继承了 LinearLayout,因此它完全可以支持 LinearLayout 所支持的全部 XML 属性。

XML 属性相关用法说明如下。

(1) android:collapseColumns 和 setColumnsCollapsed(int,boolean)方法用来设置需要隐藏的列的序列号,多个用逗号隔开。

(2) android:shrinkColumns 和 setShrinkAllColumns(boolean)方法用来设置被收缩的列的序列号,多个用逗号隔开。

(3) android:stretchColumns 和 setStretchAllColumnds(boolean)方法用来设置允许被拉伸的列的序列号,多个用逗号隔开。

android:stretchColumns 和 android:shrinkColumns 这两个属性是 TableLayout 所特有的,stretchColumns 设置可伸展的列,该列可以向行方向伸展,最多可占据一整行。shrinkColumns 设置可收缩的列,当该列子控件的内容太多,已经挤满所在行,那么该子控件的内容将往列方向显示。TableLayout 还有一个属性 android:collapseColumns,用来隐藏列。例如 android:stretchColumns="0"表示第 0 列可伸展,android:shrinkColumns="1,2",表示第 1、2 列皆可收缩,android:collapseColumns="0",表示第 0 列被隐藏。

表格布局的子对象不能指定 android:layout_width 属性,宽度只能是"match_parent"。不过子对象可以定义 android:layout_height 属性,其默认值是 wrap_content。如果子对象是 TableRow,其高度一直是 wrap_content。

在 Andriod 中,在 XML 布局文件中定义表格布局管理器可以使用<TableLayout>标记,格式如下。

```
<TableLayout
属性列表>
<!--添加需要的控件-->
</TableLayout>
```

【例 5-3】 表格布局管理器的应用。

实现步骤如下。

新建一个模块,名字为 ch05_3,打开项目文件夹中 res\layout 目录下的 activity_main.xml 文件,将其中的代码替换为以下代码。

```
<?xml version="1.0" encoding="utf-8"?>
<TableLayout xmlns:android="http://schemas.android.com/apk/res/android"
    android:layout_width="match_parent"
    android:layout_height="match_parent"
    android:background="#e8eded"
    android:stretchColumns="0,1,2"
>
<TableRow>
<TextView
        android:layout_gravity="center"
```

```xml
            android:text="姓名"/>
    <TextView
            android:layout_gravity="center"
            android:text="学校"/>
    <TextView
            android:layout_gravity="center"
            android:text="专业"/>
</TableRow>
<View
            android:layout_width="match_parent"
            android:layout_height="3dip"
            android:background="#ff909090"/>
<TableRow >
<TextView
            android:layout_gravity="center"
            android:text="张三"/>
<TextView
            android:layout_gravity="center"
            android:text="北京医科大学"/>
<TextView
            android:layout_gravity="center"
            android:text="临床医学"/>
</TableRow>
<TableRow >
<TextView
            android:layout_gravity="center"
            android:text="李四"/>
<TextView
            android:layout_gravity="center"
            android:text="大连东软信息学院"/>
<TextView
            android:layout_gravity="center"
            android:text="计算机系"/>
</TableRow>
</TableLayout>
```

在以上代码中，一个 TableLayout 布局包含了多个 TableRow，具体包含关系如图 5.5 所示。

程序运行结果如图 5.6 所示。

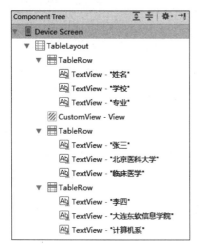

图 5.5　表格布局管理器布局结构图

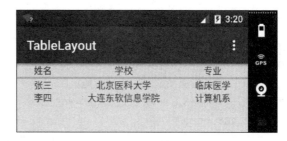

图 5.6　表格布局管理器运行结果图

5.4　网格布局管理器

网格布局管理器(GridLayout)是 Android 4.0 后新支持的布局方式,在界面设计上比表格布局更加灵活。在网格布局中,界面元素可以占用多个网格。而在表格布局中无法实现,只能将元素指定在一个表格行中,不能跨越多个表格行。表格布局中行的高度和列的宽度,完全取决于本行或本列中高度最高或宽度最宽的界面元素。

在 Android 中,在 XML 布局文件中定义帧布局管理器可以使用<GridLayout>标记,格式如下。

```
<GridLayout
属性列表>
<!-添加需要的控件-->
</ GridLayout >
```

下面通过一个案例来讲解网格布局管理器的应用。

【例 5-4】　网格布局管理器的应用。

实现步骤如下。

新建一个模块,名字为 ch05_4,打开项目文件夹中 res\layout 目录下的 activity_main.xml 文件,将其中的代码替换为以下代码。

```
<?xml version="1.0" encoding="utf-8"?>
<GridLayout xmlns:android="http://schemas.android.com/apk/res/android"
    android:layout_width="match_parent"
    android:layout_height="match_parent"
    android:layout_column="4">
  <TextView
    android:layout_width="match_parent"
```

```xml
        android:layout_height="wrap_content"
        android:layout_columnSpan="4"
        android:text="这是 GridLayout 示例"
        android:gravity="center"
        android:id="@+id/textView" />
    <TextView
        android:layout_width="wrap_content"
        android:layout_height="wrap_content"
        android:text="用户名"
        android:id="@+id/textView2"
        android:layout_row="1"
        android:layout_column="0" />
    <EditText
        android:layout_width="wrap_content"
        android:layout_height="wrap_content"
        android:id="@+id/editText"
        android:layout_row="1"
        android:layout_columnSpan="2"
        android:ems="8"/>
    <TextView
        android:layout_width="wrap_content"
        android:layout_height="wrap_content"
        android:text="密码"
        android:id="@+id/textView3"
        android:layout_row="2"
        android:layout_column="0" />
    <EditText
        android:layout_width="wrap_content"
        android:layout_height="wrap_content"
        android:id="@+id/editText2"
        android:layout_row="2"
        android:layout_columnSpan="2"
        android:ems="8"/>
    <Button
        android:layout_width="wrap_content"
        android:layout_height="wrap_content"
        android:text="清空输入"
        android:id="@+id/button"
        android:layout_row="3"
        android:layout_column="1" />
    <Button
        android:layout_width="wrap_content"
        android:layout_height="wrap_content"
        android:text="下一步"
```

```
        android:id="@+id/button5"
        android:layout_row="3"
        android:layout_column="2" />
</GridLayout>
```

案例的布局结构如图 5.7 所示。

由图 5.7 所示结构图可见,本案例中应用的是网格布局管理器,网格的行和列并没有设置为固定的值,只设置了水平排列的方式。网格布局中共添加了七个组件,这些组件最后以什么样的结果运行,是通过在 XML 文件中的属性设置实现的。例如,对显示"用户名"的 TextView 控件进行了如下设置。

```
android:layout_row="1"
android:layout_column="0"
```

这两个属性表示设置了行的值为 1 和列的值为 0。

另外,对显示"这是 GridLayout 示例"的 TextView 控件设置了如下属性。

```
android:layout_columnSpan="4"
```

表示列的跨度为"4"。正是通过对属性的灵活设置,实现了网格布局管理器的灵活应用,从而实现用户的各类需求。

最后来看一下程序运行效果,如图 5.8 所示。

图 5.7 网格布局管理器布局结构图

图 5.8 网格布局管理器运行效果图

5.5 帧布局管理器

FrameLayout 表示帧布局管理器。在帧布局管理器中,每加入一个控件都会创建一个空白区域,通常称为一帧,这些帧会根据 gravity 属性执行自动对齐。默认情况下,帧布局从屏幕的左上角(0,0)坐标点开始布局,多个控件层叠排序,后面的控件覆盖前面的控件。

此布局可以放置多个 view,但只有一个 view 可以显示,通常使用此布局处理在同一位置不同情况下显示不同内容的控件。

在 Andriod 中,在 XML 布局文件中定义帧布局管理器可以使用<FrameLayout>标记,格式如下。

```
<FrameLayout
属性列表>
<!--添加需要的控件-->
</FrameLayout>
```

【例 5-5】 帧布局管理器的应用。

实现步骤如下。

新建一个模块,名字为 ch05_5,打开项目文件夹中 res\layout 目录下的 activity_main.xml 文件,将其中的代码替换为以下代码。

```
<?xml version="1.0" encoding="utf-8"?>
<LinearLayout xmlns:android="http://schemas.android.com/apk/res/android"
    android:layout_width="fill_parent"
    android:layout_height="fill_parent"
    android:orientation="vertical"
    android:background="#100f0f"
    >
    <LinearLayout
        android:layout_width="fill_parent"
        android:layout_height="wrap_content"
        android:orientation="horizontal"
        >
        <Button
            android:layout_width="wrap_content"
            android:layout_height="wrap_content"
            android:text="FrameA"
            android:id="@+id/btnA"
            />
        <Button
            android:layout_width="wrap_content"
            android:layout_height="wrap_content"
            android:text="FrameB"
            android:id="@+id/btnB"
            />
        <Button
            android:layout_width="wrap_content"
            android:layout_height="wrap_content"
            android:text="FrameC"
            android:id="@+id/btnC"
            />

    </LinearLayout>
    <FrameLayout
```

```xml
    android:layout_width="match_parent"
    android:layout_height="match_parent">

    <LinearLayout
        android:layout_width="match_parent"
        android:layout_height="match_parent"
        android:orientation="vertical"
        android:visibility="visible"
        android:background="#aaaaaaff"
        android:id="@+id/frameA">

        <TextView
            android:id="@+id/textView1"
            android:layout_width="match_parent"
            android:layout_height="match_parent"
            android:text="This is FrameA"
            android:textSize="42px"/>
    </LinearLayout>
    <LinearLayout
        android:layout_width="match_parent"
        android:layout_height="match_parent"
        android:orientation="vertical"
        android:id="@+id/frameB"
        android:background="#aa00ffaa">
        <TextView
            android:id="@+id/textView2"
            android:layout_width="match_parent"
            android:layout_height="match_parent"
            android:text="This is FrameB"
            android:textSize="42px"/>
    </LinearLayout>
    <LinearLayout
        android:layout_width="match_parent"
        android:layout_height="match_parent"
        android:orientation="vertical"
        android:id="@+id/frameC"
        android:background="#aa00ffaa">
        <TextView
            android:id="@+id/textView3"
            android:layout_width="match_parent"
            android:layout_height="match_parent"
            android:text="This is FrameC"
            android:textSize="42px"/>
    </LinearLayout>
```

```
        </FrameLayout>
</LinearLayout>
```

程序的布局结构如图 5.9 所示。

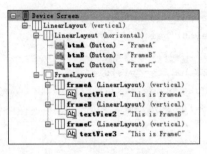

图 5.9 帧布局管理器布局结构图

在 MainActivity.java 文件中添加如下代码,让按钮能够进行事件处理,从而显示帧布局切换的效果。

```java
import android.os.Bundle;
import android.support.v7.app.AppCompatActivity;
import android.view.View;
import android.view.View.OnClickListener;
import android.widget.Button;
import android.widget.LinearLayout;

public class MainActivity extends AppCompatActivity {
    private Button btnA,btnB,btnC;
    private LinearLayout viewA,viewB,viewC;
    @Override
    protected void onCreate(Bundle savedInstanceState){
        super.onCreate(savedInstanceState);
        setContentView(R.layout.activity_main);
            btnA=(Button)this.findViewById(R.id.btnA);
        btnB=(Button)this.findViewById(R.id.btnB);
        btnC=(Button)this.findViewById(R.id.btnC);
        viewA=(LinearLayout)this.findViewById(R.id.frameA);
        viewB=(LinearLayout)this.findViewById(R.id.frameB);
        viewC=(LinearLayout)this.findViewById(R.id.frameC);
        btnA.setOnClickListener(new OnClickListener(){

            @Override
            public void onClick(View v){
                viewA.setVisibility(View.VISIBLE);
                viewB.setVisibility(View.GONE);
                viewC.setVisibility(View.GONE);
```

```
            }
        });
        btnB.setOnClickListener(new OnClickListener(){
            
            @Override
            public void onClick(View v){
                viewA.setVisibility(View.GONE);
                viewB.setVisibility(View.VISIBLE);
                viewC.setVisibility(View.GONE);
            }
        });
        btnC.setOnClickListener(new OnClickListener(){
            @Override
            public void onClick(View v){
                viewA.setVisibility(View.GONE);
                viewB.setVisibility(View.GONE);
                viewC.setVisibility(View.VISIBLE);
                
            }
        });
    }
}
```

单击按钮 FrameA,显示如图 5.10 所示的界面。

单击按钮 FrameC,显示如图 5.11 所示的界面。

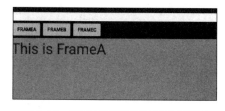
图 5.10　帧布局管理器运行结果图 1

图 5.11　帧布局管理器运行结果图 2

5.6　向容器中手动添加控件

Android 程序的需求越来越复杂,编写程序时,经常需要动态向程序中添加组件。前面布局管理器中的组件都是直接布局好的,要在布局容器中动态添加控件,使用的方法如下。

(1) addView。

添加控件到布局容器。

(2) removeView。

在布局容器中删掉已有的控件。

接下来以动态添加 Button 控件为例学习具体的方法,添加其他控件也是同样的道理。

【例 5-6】 手动添加控件。

实现步骤如下。

新建一个模块,名字为 ch05_6,打开项目文件夹中 res\layout 目录下的 activity_main.xml 文件,将其中的代码替换为以下代码。

```
<?xml version="1.0" encoding="utf-8"?>
<LinearLayout xmlns:android="http://schemas.android.com/apk/res/android"
    android:orientation="vertical" android:layout_width="match_parent"
    android:layout_height="match_parent"
    android:id="@+id/container">
<EditText
    android:layout_width="match_parent"
    android:layout_height="wrap_content"
    android:id="@+id/editText" />
<Button
    android:layout_width="wrap_content"
    android:layout_height="wrap_content"
    android:text="添加控件"
    android:id="@+id/button"
    android:layout_gravity="center_horizontal" />
</LinearLayout>
```

在 MainActivity.java 中添加如下语句。

```
EditText edt=null;
Button btn=null;
LinearLayout c=null;
@Override
protected void onCreate(Bundle savedInstanceState){
    btn=(Button)findViewById(R.id.button);
    edt=(EditText)findViewById(R.id.editText);
    c=(LinearLayout)findViewById(R.id.container);
    btn.setOnClickListener(new View.OnClickListener(){
@Override
        public void onClick(View arg0){
            Button newbtn=new Button(arg0.getContext());
            newbtn.setText(edt.getText());
            c.addView(newbtn);
        }
    });
```

程序的界面布局如图 5.12 所示。

程序运行结果如图 5.13 所示。

图 5.12　手动添加控件布局结构图　　　图 5.13　手动添加组件运行结果图

5.7　项目实战：CoffeeStore 首页的界面开发

5.7.1　项目分析

本部分要实现项目中首页的界面开发。完成界面开发时,首先要确定界面的布局,即确定界面所用的布局和布局间的嵌套关系。

首先来看首页页面的组成,首页界面涉及多个布局管理器的应用。最外层是 LinearLayout 布局管理器,以垂直的方式布局,内层首先嵌套一个 FrameLayout 布局管理器,以实现将来在程序中添加广告的展示,当前只留出位置,内容在后续章节添加。这部分的布局如图 5.14 所示。

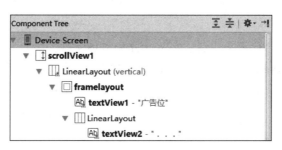

图 5.14　CoffeeStore 首页布局结构图 1

中间以 TableLayout 布局管理器和 LinearLayout 布局管理器两种布局管理器来布局,显示各类功能。图 5.15 中(a)图表示表格布局中第一行的布局,(b)图表示第二行的布局,由图可见,两部分的组成是相同的。

接下来是第三部分,将来是热门店铺的推荐、热门商品的推荐等,布局结构如图 5.16 所示。这个版块也是应用线性布局管理器来实现的,应用到的其他控件将在后续章节介绍,这里不赘述。

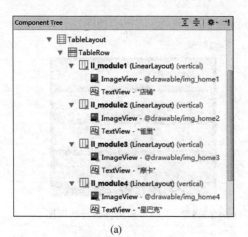 

图 5.15 CoffeeStore 首页布局结构图 2

图 5.16 CoffeeStore 首页布局结构图 3

5.7.2 项目实现

接下来就通过程序代码来讲解这部分功能的具体实现。

页面布局文件 fragment_home.xml 的代码如下。

```xml
<?xml version="1.0" encoding="utf-8"?>
<ScrollView xmlns:android="http://schemas.android.com/apk/res/android"
    android:id="@+id/scrollView1"
    android:layout_width="match_parent"
    android:layout_height="match_parent">

    <LinearLayout
        android:layout_width="match_parent"
        android:layout_height="match_parent"
        android:orientation="vertical">
        <FrameLayout
            android:id="@+id/framelayout"
            android:layout_width="match_parent"
            android:layout_height="150dp"
```

```xml
        android:background="#00ff00">
        <TextView
            android:id="@+id/textView1"
            android:layout_width="wrap_content"
            android:layout_height="wrap_content"
            android:layout_gravity="center"
            android:text="广告位"
            android:textSize="36sp" />
        <LinearLayout
            android:layout_width="match_parent"
            android:layout_height="wrap_content"
            android:layout_gravity="bottom"
            android:background="#88252525"
            android:gravity="center"
            android:padding="3dp">
            <TextView
                android:id="@+id/textView2"
                android:layout_width="wrap_content"
                android:layout_height="wrap_content"
                android:text="..."
                android:textColor="#ff0000"
                android:textSize="24sp" />
        </LinearLayout>
    </FrameLayout>
    <GridLayout
        android:layout_width="match_parent"
        android:layout_height="match_parent"
        android:layout_weight="1"
        android:background="#fafafa"
        android:columnCount="4"
        android:rowCount="2">
        <LinearLayout
            android:id="@+id/ll_module1"
            android:layout_width="wrap_content"
            android:layout_height="wrap_content"
            android:layout_columnWeight="1"
            android:layout_rowWeight="1"
            android:orientation="vertical">
            <ImageView
                android:layout_width=" wrap_content "
                android:layout_height="50dp"
                android:layout_gravity="center"
                android:padding="5dp"
                android:src="@drawable/img_home1" />
```

```xml
<TextView
    android:layout_width="match_parent"
    android:layout_height="wrap_content"
    android:gravity="center_horizontal"
    android:text="店铺"
    android:textColor="#929292"
    android:textSize="13sp" />
</LinearLayout>
<LinearLayout
    android:id="@+id/ll_module2"
    android:layout_width="wrap_content"
    android:layout_height="wrap_content"
    android:layout_columnWeight="1"
    android:layout_rowWeight="1"
    android:orientation="vertical">
    <ImageView
        android:layout_width=" wrap_content "
        android:layout_height="50dp"
        android:layout_gravity="center"
        android:padding="5dp"
        android:src="@drawable/img_home2" />
    <TextView
        android:layout_width="match_parent"
        android:layout_height="wrap_content"
        android:gravity="center_horizontal"
        android:text="雀巢"
        android:textColor="#929292"
        android:textSize="13sp" />
</LinearLayout>
<LinearLayout
    android:id="@+id/ll_module3"
    android:layout_width="wrap_content"
    android:layout_height="wrap_content"
    android:layout_columnWeight="1"
    android:layout_rowWeight="1"
    android:orientation="vertical">
    <ImageView
        android:layout_width=" wrap_content p"
        android:layout_height="50dp"
        android:layout_gravity="center"
        android:padding="5dp"
        android:src="@drawable/img_home3" />
    <TextView
        android:layout_width="match_parent"
```

```xml
            android:layout_height="wrap_content"
            android:gravity="center_horizontal"
            android:text="摩卡"
            android:textColor="#929292"
            android:textSize="13sp" />
    </LinearLayout>
    <LinearLayout
        android:id="@+id/ll_module4"
        android:layout_width="wrap_content"
        android:layout_height="wrap_content"
        android:layout_columnWeight="1"
        android:layout_rowWeight="1"
        android:orientation="vertical">
        <ImageView
            android:layout_width=" wrap_content p"
            android:layout_height="50dp"
            android:layout_gravity="center"
            android:padding="5dp"
            android:src="@drawable/img_home4" />
        <TextView
            android:layout_width="match_parent"
            android:layout_height="wrap_content"
            android:gravity="center_horizontal"
            android:text="星巴克"
            android:textColor="#929292"
            android:textSize="13sp" />
    </LinearLayout>
    <LinearLayout
        android:id="@+id/ll_module5"
        android:layout_width="wrap_content"
        android:layout_height="wrap_content"
        android:layout_columnWeight="1"
        android:layout_rowWeight="1"
        android:orientation="vertical">
        <ImageView
            android:layout_width=" wrap_content p"
            android:layout_height="50dp"
            android:layout_gravity="center"
            android:padding="5dp"
            android:src="@drawable/img_home5" />
        <TextView
            android:layout_width="match_parent"
            android:layout_height="wrap_content"
            android:gravity="center_horizontal"
```

```xml
            android:text="优惠券"
            android:textColor="#929292"
            android:textSize="13sp" />
    </LinearLayout>
    <LinearLayout
        android:id="@+id/ll_module6"
        android:layout_width="wrap_content"
        android:layout_height="wrap_content"
        android:layout_columnWeight="1"
        android:layout_rowWeight="1"
        android:orientation="vertical">
        <ImageView
            android:layout_width=" wrap_content "
            android:layout_height="50dp"
            android:layout_gravity="center"
            android:padding="5dp"
            android:src="@drawable/img_home6" />
        <TextView
            android:layout_width="match_parent"
            android:layout_height="wrap_content"
            android:gravity="center_horizontal"
            android:text="我的关注"
            android:textColor="#929292"
            android:textSize="13sp" />
    </LinearLayout>
    <LinearLayout
        android:id="@+id/ll_module7"
        android:layout_width="wrap_content"
        android:layout_height="wrap_content"
        android:layout_columnWeight="1"
        android:layout_rowWeight="1"
        android:orientation="vertical">
        <ImageView
            android:layout_width=" wrap_content "
            android:layout_height="50dp"
            android:layout_gravity="center"
            android:padding="5dp"
            android:src="@drawable/img_home7" />
        <TextView
            android:layout_width="match_parent"
            android:layout_height="wrap_content"
            android:gravity="center_horizontal"
            android:text="摇大奖"
            android:textColor="#929292"
```

```xml
                android:textSize="13sp" />
        </LinearLayout>
        <LinearLayout
            android:id="@+id/ll_module8"
            android:layout_width="wrap_content"
            android:layout_height="wrap_content"
            android:layout_columnWeight="1"
            android:layout_rowWeight="1"
            android:orientation="vertical">
            <ImageView
                android:layout_width=" wrap_content "
                android:layout_height="50dp"
                android:layout_gravity="center"
                android:padding="5dp"
                android:src="@drawable/img_home8" />
            <TextView
                android:layout_width="match_parent"
                android:layout_height="wrap_content"
                android:gravity="center_horizontal"
                android:text="物流查询"
                android:textColor="#929292"
                android:textSize="13sp" />
        </LinearLayout>
    </GridLayout>
    <View
        android:layout_width="match_parent"
        android:layout_height="4dp"
        android:background="#ff0000" />
    <LinearLayout
        android:id="@+id/recommand"
        android:layout_width="match_parent"
        android:layout_height="150dp"
        android:orientation="vertical"
        android:scrollbars="none">
        <TextView
            android:layout_width="match_parent"
            android:layout_height="wrap_content"
            android:gravity="center_horizontal"
            android:text="推荐商品"
            android:textColor="#929292"
            android:textSize="15sp" />
        <GridView
            android:id="@+id/grid"
            android:layout_width="match_parent"
```

```xml
                    android:layout_height="wrap_content"
                    android:numColumns="2" />"
            </LinearLayout>
            <View
                android:layout_width="match_parent"
                android:layout_height="4dp"
                android:background="#ff0000" />
            <LinearLayout
                android:id="@+id/discount"
                android:layout_width="match_parent"
                android:layout_height="150dp"
                android:background="#6666ff"
                android:orientation="vertical">
                <TextView
                    android:layout_width="match_parent"
                    android:layout_height="wrap_content"
                    android:gravity="center_horizontal"
                    android:text="打折商品"
                    android:textColor="#929292"
                    android:textSize="13sp" />
            </LinearLayout>
            <View
                android:layout_width="match_parent"
                android:layout_height="4dp"
                android:background="#ff0000" />
            <LinearLayout
                android:layout_width="match_parent"
                android:layout_height="150dp"
                android:background="#ff6666"
                android:orientation="vertical"
                android:scrollbars="none">
                <TextView
                    android:layout_width="match_parent"
                    android:layout_height="wrap_content"
                    android:gravity="center_horizontal"
                    android:text="热销商品"
                    android:textColor="#929292"
                    android:textSize="13sp" />
            </LinearLayout>
        </LinearLayout>
</ScrollView>
```

5.7.3 项目说明

本项目首页涵盖的内容较多,布局复杂,因此首页的设计用到了多种布局。先把首页分成上、中、下三块,再以此进行每块的设计和布局,最终实现总体效果,运行 Activity 文件显示的界面效果如图 5.17 所示。

图 5.17　CoffeeStore 首页运行效果图

本 章 小 结

　　布局管理器的作用是根据屏幕大小管理容器内的控件,自动适配组件在手机屏幕中的位置。本章共讲述 5 种布局管理器,各有特点。
　　(1) 线性布局管理器会将容器中的组件一个一个排列起来,LinearLayout 可以通过 android:orientation 属性控制组件横向或者纵向排列。
　　(2) 在相对布局管理器中,子组件的位置总是相对兄弟组件、父容器来决定。
　　(3) 表格布局管理器采用行、列的形式管理子组件,但是并不需要声明有多少行和列,只需要添加 TableRow 和组件就可以控制表格的行数和列数,这一点与网格布局有所不同,网格布局需要指定行列数。
　　(4) 网格布局管理器将整个容器划分成 rows×columns 个网格,每个网格可以放置一个组件,还可以设置一个组件横跨多少列和多少行,不存在一个网格放多个组件的情况。
　　(5) 帧布局管理器为每个组件创建一个空白区域,一个区域成为一帧,这些帧会根据 FrameLayout 中定义的 gravity 属性自动对齐。
　　本章最后介绍了向容器中手动添加控件的方法。

本 章 习 题

1. 如何设置线性布局为水平排列？
2. 表格布局和网格布局的区别是什么？
3. 如何应用帧布局管理器？
4. 举例说明 android:gravity 与 android:layout_gravity 的区别。
5. 请阐述 GridLayout 的使用场合及用法。
6. Android 中常用的五个布局有哪些？
7. 利用本章的知识，设计完成项目中的注册功能界面。用户注册时需要填写用户名、密码、密码确认等信息，请采用合适的布局管理器实现。
8. 利用本章的知识，完成项目中评论管理界面的设计，界面实现用户可以填写评论，填写后提交评论的功能，请采用合适的布局管理器实现。
9. 本章的项目中用到了 TableLayout 布局管理器实现界面，请改写为使用 GridLayout 布局管理器实现，分析效果有何区别。
10. 编写 Android 程序，利用所学的布局管理器实现图 5.18 所示界面。
11. 利用所学的布局方法实现图 5.19 所示用户界面。

图 5.18 习题 10 实现界面

图 5.19 习题 11 实现界面

12. 使用适合的布局方法实现图 5.20 所示用户界面。

图 5.20 习题 12 实现界面

Android 基本控件

本章概述

通过本章的学习,掌握 Android 基本控件的使用,为实现 App 功能提供技术支持。本章讲授的控件应用,包括文本类控件、按钮类控件、日期和时间类控件、进度条和滑动条等控件,从控件的创建、属性和方法的使用等角度讲解基本控件,通过实例强化控件的使用。

学习重点与难点

重点:

(1) 文本类控件。

(2) 按钮类控件。

(3) 日期和时间类控件。

(4) 进度条控件。

难点:

(1) 控件的事件处理。

(2) 控件的综合运用。

学习建议

读者在学习中要多动手实践,多查阅资料,广纳精华,多练习思考本教材中提供的项目,从而更加熟练地掌握控件的综合应用,以及控件事件处理方式的应用。

6.1 文本类控件

6.1.1 TextView

Android 中的 TextView 表示文本框,用于在屏幕上显示文本,主要显示不可编辑的文本。它与 Java 中的文本框控件不同,它相当于 Java 中的标签,TextView 可以自动识别并链接如 E-mail 地址、Web 地址、电话号码等特殊字符,它包含的主要属性如表 6.1 所示。

表 6.1　TextView 常用属性说明

属　　性	说　　明
android:text	设置 TextView 显示的文本内容
android:textColor	设置文本颜色
android:textSize	设置文本字体的大小，支持度量单位：px(像素)/dp/sp/in/mm
android:textStyle	设置文本字体的风格，取值一般为 bold、italic 等
android:height	设置文本区域的高度，支持度量单位：px(像素)/dp/sp/in/mm
android:layout_height	设置文本相对父类的长度的值，取值一般为 match_parent、wrap_content、match_parent
android:width	设置文本区域的宽度，支持度量单位：px(像素)/dp/sp/in/mm，如果 android:layout_width="match_parent"，那么设置 android:width 是没有意义的
android:layout_width	设置文本相对父类的宽度的值，取值一般为 match_parent、wrap_content、match_parent
android:layout_gravity	设置该控件相对于父 view 的位置
android:numeric	如果被设置，该 TextView 有一个数字输入法。有如下值设置：integer 正整数、signed 带符号整数、decimal 带小数点浮点数
android:password	以小点"."显示文本
android:phoneNumber	设置为电话号码的输入方式
android:singleLine	设置单行显示。如果和 layout_width 一起使用，当文本不能全部显示时，后面用"…"来表示。如果不设置 SingleLine 或者设置为 false，文本将自动换行

属性还有很多，如果需要，可自行查阅帮助文档，这里不再赘述。

在 XML 文件中定义 TextView 的格式如下。

```
<TextView
属性设置
    />
```

下面通过例子看一下 TextView 的应用。

【例 6-1】　TextView 控件的应用。

本案例介绍了 TextView 控件的各种显示效果，包括文字居中、文字跑马灯效果、设置文字阴影效果、设置网址超链接效果、设置文字超链接效果、设置电话超链接效果、设置字形、设置文字缩放、设置行间距、设置行间距的倍数等内容，具体的应用主要是通过 TextView 控件的属性进行设置。

实现步骤如下。

新建一个模块，名字为 ch06_1，打开项目文件夹中 res\layout 目录下的 activity_main.xml 文件，将其中的代码替换为以下代码。

```
<?xml version="1.0" encoding="utf-8"?>
```

```xml
<LinearLayout xmlns:android="http://schemas.android.com/apk/res/android"
    android:orientation="vertical"
    android:layout_width="match_parent"
    android:layout_height="match_parent"
    android:background="#ffF4F4F4"
    android:padding="8px">

<!--文字居中 -->
<TextView android:id="@+id/testGravity"
    android:layout_width="match_parent"
    android:layout_height="wrap_content"
    android:text="居中文字"
    android:gravity="center"
    android:background="#ff00ff00"
    android:textColor="#ff000000"
    android:textSize="18sp" />
<!--文字跑马灯效果 -->
<TextView android:id="@+id/testEllipsize"
    android:layout_width="100px"
    android:layout_height="wrap_content"
    android:text="跑马灯文字效果"
    android:ellipsize="marquee"
    android:marqueeRepeatLimit="marquee_forever"
    android:singleLine="true"
    android:focusable="true"
    android:focusableInTouchMode="true"
    android:textColor="#ff000000"
    android:textSize="20sp" />
<!--设置文字阴影效果 -->
<TextView android:id="@+id/testShadow" android:layout_width="match_parent"
    android:layout_height="wrap_content" android:text="文字阴影效果"
    android:textColor="#ff000000" android:shadowColor="#ff0000"
    android:shadowRadius="3.0" />
<!--设置网址超链接效果 -->
<TextView android:id="@+id/testAutoLink1"
    android:layout_width="match_parent"
    android:layout_height="wrap_content"
    android:text="网址超链接:http://www.baidu.com" android:textColor=
    "#ff000000"
    android:autoLink="web"
    android:textColorLink="#ff0000ff" />
<!--设置文字超链接效果 -->
<TextView android:id="@+id/testAutoLink2"
    android:layout_width="match_parent"
    android:layout_height="wrap_content"
    android:text="@string/txtlink"
    android:textColor="#ff000000" />
```

```xml
<!--设置电话超链接效果 -->
<TextView android:id="@+id/testAutoLink2"
    android:layout_width="match_parent" android:layout_height="wrap_content"
    android:text="@string/tellink" android:textColor="#ff000000"
    android:autoLink="phone" />
<!--设置字形 -->
<TextView android:id="@+id/testTextStyle"
    android:layout_width="match_parent"
    android:layout_height="wrap_content"
    android:text="斜体"
    android:textColor="#ff000000"
    android:textStyle="italic" />
<!--设置文字缩放 -->
<TextView android:id="@+id/testTextScaleX"
    android:layout_width="match_parent"
    android:layout_height="wrap_content"
    android:text="hello 0.5f"
    android:textColor="#ff000000"
    android:textScaleX="0.5" />
<TextView android:layout_width="match_parent"
    android:layout_height="wrap_content"
    android:text="hello 1.0f"
    android:textColor="#ff000000"
    android:textScaleX="1.0" />
<TextView android:layout_width="match_parent"
    android:layout_height="wrap_content"
    android:text="hello 1.5f"
    android:textColor="#ff000000"
    android:textScaleX="1.5" />
<TextView android:layout_width="match_parent"
    android:layout_height="wrap_content"
    android:text="hello 2.0f"
    android:textColor="#ff000000"
    android:textScaleX="2.0" />
<TextView android:layout_width="match_parent"
    android:layout_height="wrap_content"
    android:text="hello 2.5f"
    android:textColor="#ff000000"
    android:textScaleX="2.5" />
<!--设置行间距 -->
<TextView android:id="@+id/testLineSpacingExtra"
    android:layout_width="match_parent"
    android:layout_height="wrap_content"
    android:text="@string/lineheight1"
    android:textColor="#ff000000"
    android:lineSpacingExtra="4px" />
```

```
<!--设置行间距的倍数-->
<TextView android:id="@+id/testLineSpacingMultiplier"
    android:layout_width="match_parent"
    android:layout_height="wrap_content"
    android:text="@string/lineheight2"
     android:textColor="#ff000000"
    android:lineSpacingMultiplier="1.5" />
</LinearLayout>
```

程序的控件布局结构如图 6.1 所示。

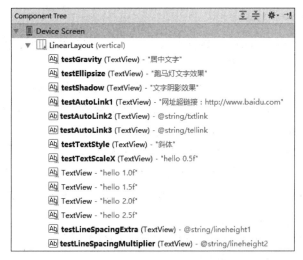

图 6.1 TextView 控件布局结构图

程序运行结果如图 6.2 所示。

图 6.2 TextView 控件运行结果图

6.1.2　AutoCompleteTextView

自动文本框用 AutoCompleteTextView 表示。用户输入一定字符后显示一个下拉菜单，供用户从中选择选项，当用户选择某个菜单项后，按所选自动填写该文本框。

在屏幕中添加自动文本框，可在 XML 布局文件中通过＜AutoCompleteTextView＞标记添加，格式如下。

```
<AutoCompleteTextView
属性列表>
</AutoCompleteTextView>
```

AutoCompleteTextView 控件是由 TextView 派生的，所以它支持 EditText 控件提供的属性，同时还支持以下 XML 属性。

android:completionHint：用于为弹出的下拉菜单指定提示标题。
android:completionThreshold：用于指定用户至少输入几个字符才会显示提示。
android:dropDownAnchor：下拉列表的锚点或挂载点。
android:dropDownHeight：用于指定下拉菜单的高度。
android:dropDownWidth：用于指定下拉菜单的宽度。
android:dropDownHorizontalOffset：用于指定下拉菜单与文本之间的水平偏移，下拉菜单默认与文本框左对齐。
android:dropDownSelector：下拉列表被选中的行的背景。
android:ems：需要编辑的字符串长度。

6.1.3　MultiAutoCompleTextView

该控件可支持选择多个值（在多次输入的情况下），分别用分隔符分开，并且在每个值选中的时候，再次输入值时会自动匹配。可用在发短信、发邮件时选择联系人这种类型当中。使用时需要执行设置分隔符方法。

MultiAutoCompleteTextView 的使用和 AutoCompleteTextView 类似，只是需要设置分隔符。

具体的使用方法为在 setAdapter()方法后添加如下内容。

```
setTokenizer(new MultiAutoCompleteTextView.CommaTokenizer());
```

6.1.4　EditText

Android 中使用 EditText 表示输入框，用于在屏幕上显示文本输入框，这与 Java 中的文本框控件功能类似。不同之处在于，Android 中的编辑框控件可以输入单行文本，也可以输入多行文本，还可以输入指定格式的文本，如电话号码、密码、E-mail 地址等内容。

在 Android 中，可使用两种方法向屏幕中添加编辑框，通常在 XML 布局文件中使用＜EditText＞标记添加，添加格式如下。

```
<EditText
属性列表>
</EditText>
```

由于 EditText 类是 TextView 的子类,所以 TextView 的属性适用于 EditText 控件。特别需要注意的是,在 EditText 控件中,android:inputType 属性可以帮助输入法显示合适的类型。例如,要添加一个密码框,可以将 android:inputType 属性设置为 textPassword。

通常,使用 EditText 控件时,还需要获取该控件输入的文本内容,可以通过编辑框控件提供的 getText()方法实现。使用该方法时,要先获取编辑框控件,然后再调用 getText()方法。例如,要获取布局文件中添加的 ID 属性为 txt 的编辑框内容,可通过如下代码实现。

```
EditText txt=(EditText)(this.findViewById(R.id.txt));
String   txtText=txt.getText().toString();
```

【例 6-2】 EditText 控件、AutoCompleteTextView 控件和 MultiAutoCompleteTextView 控件的应用。

本案例介绍了 3 个控件的应用,使用 EditText 控件的输入文本功能,通过对 EditText 控件进行事件处理实现了对脏话的过滤功能,同时也实现了用 AutoCompleteTextView 和 MultiAutoCompleteTextView 两个控件自动完成单个和多个文本的输入功能。

实现步骤如下。

新建一个模块,名字为 ch06_2,打开项目文件夹中 res\layout 目录下的 activity_main.xml 文件,将其中的代码替换为以下代码。

```
<?xml version="1.0" encoding="utf-8"?>
<LinearLayout xmlns:android="http://schemas.android.com/apk/res/android"
    android:layout_width="match_parent"
    android:layout_height="match_parent"
    android:orientation="vertical" >

<AutoCompleteTextView
        android:id="@+id/autoCompleteTextView1"
        android:layout_width="match_parent"
        android:layout_height="wrap_content"
        android:ems="10"
        android:text="AutoCompleteTextView"
         />

<MultiAutoCompleteTextView
        android:id="@+id/multiAutoCompleteTextView1"
        android:layout_width="match_parent"
```

```xml
        android:layout_height="wrap_content"
        android:ems="10"
        android:text="MultiAutoCompleteTextView"
         />

<EditText
        android:id="@+id/edit"
        android:layout_width="match_parent"
        android:layout_height="wrap_content"
         android:text="EditText">
<!--android:ems="10"
        android:inputType="textImeMultiLine"
        android:imeOptions="actionGo"
        android:hint="只能输入数字及小写字母"
        android:digits="0123456789abcdefghijklmnopqrstuvwxyz"
        android:maxLength="5">-->
<requestFocus />
</EditText>
</LinearLayout>
```

MainActivity 中的主要代码如下。

```java
public class MainActivity extends Activity {
    private AutoCompleteTextView atv;
    private MultiAutoCompleteTextView matv;
    private EditText edittext;

    @Override
    protected void onCreate(Bundle savedInstanceState){
        //TODO Auto-generated method stub
        super.onCreate(savedInstanceState);
        setContentView(R.layout.autocomplete_text);

        edittext=(EditText)this.findViewById(R.id.edit);
        atv=(AutoCompleteTextView)this.findViewById(R.id.autoCompleteTextView1);
        matv=(MultiAutoCompleteTextView)this.findViewById(R.id.multiAutoCompleteTextView1);
        String[] items={"Baiyunshan", "Beigin", "Bengbu", "Baicheng"};

        ArrayAdapter<String> adapter=new ArrayAdapter<String>(MainActivity.this,android.R.layout.simple_dropdown_item_1line, items);
        atv.setAdapter(adapter);
        matv.setAdapter(adapter);
        matv.setTokenizer(new MultiAutoCompleteTextView.CommaTokenizer());
```

```java
edittext.addTextChangedListener(new TextWatcher(){
    private String oldText;
    private final String[] filters={"NND","TMD","草泥马"};
    
    @Override
    public void onTextChanged(CharSequence arg0, int arg1, int arg2, int arg3){
        //TODO Auto-generated method stub
        
    }
    
    @Override
    public void beforeTextChanged(CharSequence arg0, int arg1, int arg2, int arg3){
        //TODO Auto-generated method stub
        oldText=edittext.getText().toString();
        
    }
    
    @Override
    public void afterTextChanged(Editable arg0){
        //TODO Auto-generated method stub
        String newText=edittext.getText().toString();
        if(oldText.equals(newText)){
            return;
        }
        for(String filter : filters){
            if(newText.indexOf(filter)!=-1){
                newText=newText.replaceAll(filter, " * * ");
            }
        }
        edittext.setText(newText);
        edittext.invalidate();
//Invalidate()函数的作用是使整个窗口客户区无效,窗口客户无效即需要重绘
        edittext.setSelection(newText.length());
        
    }
});
}
}
```

程序的控件布局结构如图6.3所示。

EditText等控件的程序运行效果如图6.4所示。

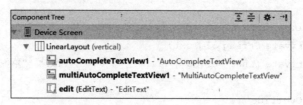

图 6.3　EditText 等控件布局结构图

在图 6.4 中的 AutoCompleteTextView 控件中输入文本"Be",程序会自动弹出两个可选项,可以选择其中一个输入,运行结果如图 6.5 所示。

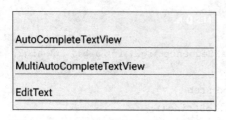

图 6.4　EditText 等控件程序运行效果图

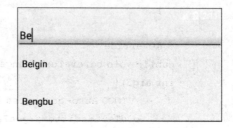

图 6.5　AutoCompleteTextView 控件程序运行结果图

在 MultiAutoCompleteTextView 控件中输入文本"Be",程序也会自动弹出两个可选项,可以选择其中一个输入,也可以选择多个,自动用","分隔,运行结果如图 6.6 所示。

EditText 控件具备文本输入的功能,本案例中设置了脏话的过滤功能。当从键盘输入"NND"时,屏幕自动显示"**",把脏话隐藏了。具体显示结果如图 6.7 所示。

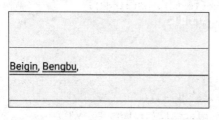

图 6.6　MultiAutoCompleteTextView
　　　　控件程序运行结果图

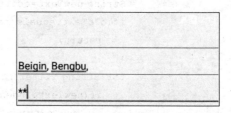

图 6.7　EditText 等控件程序运行结果图

6.2　ScrollView

ScrollView 滚动视图是指当拥有很多内容,屏幕显示不完时,需要通过滚动条来显示的视图。ScrollView 只支持垂直滚动。

具体的应用请看下面的例子。

【例 6-3】　ScrollView 控件的应用。

本案例介绍了 ScrollView 控件的应用。首先在布局文件中添加一个 ScrollView 控件,设置的布局管理器为线性布局,对齐方式为垂直的对齐方式,然后在 ScrollViewActivity 中采用手动添加组件的方式向 ScrollView 控件中添加若干个按钮控件。在程序的运行结果中可以见到,按钮的个数超出了屏幕的显示范围,因此出现了滚动条,可以垂直滚动,显示下面的组件。

实现步骤如下。

新建一个模块,名字为 ch06_3,打开项目文件夹中 res\layout 目录下的 activity_main.xml 文件,将其中的代码替换为以下代码。

```xml
<?xml version="1.0" encoding="utf-8"?>
<ScrollView xmlns:android="http://schemas.android.com/apk/res/android"
    android:layout_width="match_parent"
    android:layout_height="match_parent" >
<LinearLayout
        android:layout_width="match_parent"
        android:layout_height="match_parent"
        android:orientation="vertical"
        android:id="@+id/linear"/>"
</ScrollView>
```

ScrollViewActivity 的代码替换如下。

```java
public class ScrollViewActivity extends Activity {
    private LinearLayout linearLayout;
    @Override
    protected void onCreate(Bundle savedInstanceState) {
        super.onCreate(savedInstanceState);
        setContentView(R.layout.scrolview_layout);

        String []data={"Java 语言程序设计","软件体系结构与架构技术","大学计算机基础","软件工程实践","Web 开发技术","人机交互设计","软件测试","移动互联网应用开发","软件工程项目实训","专业导引与职涯规划","面向对象分析与编程","JavaScript 程序设计","Android 高级开发技术","算法分析与设计","面向对象系统分析与设计","Internet 技术基础","软件质量保证与测试","HTML 5 移动 Web 开发","计算机导论","自动化测试工具","软件工程概论","软件项目管理","跨平台移动应用开发","轻松玩转 Office"};
        linearLayout=(LinearLayout)this.findViewById(R.id.linear);
        for(int i=0;i<data.length;i++){
          Button btn=new Button(this);
          btn.setAllCaps(false);
          btn.setText(data[i]);
          linearLayout.addView(btn);
        }
    }
}
```

本案例的布局结构如图 6.8 所示。

程序运行结果如图 6.9 所示。

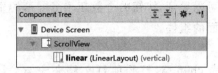

图 6.8　ScrollView 案例布局结构图

图 6.9 中并没有显示所有的程序运行结果，需要滑动滚动条，才能够看见图 6.10 所示的其他结果。

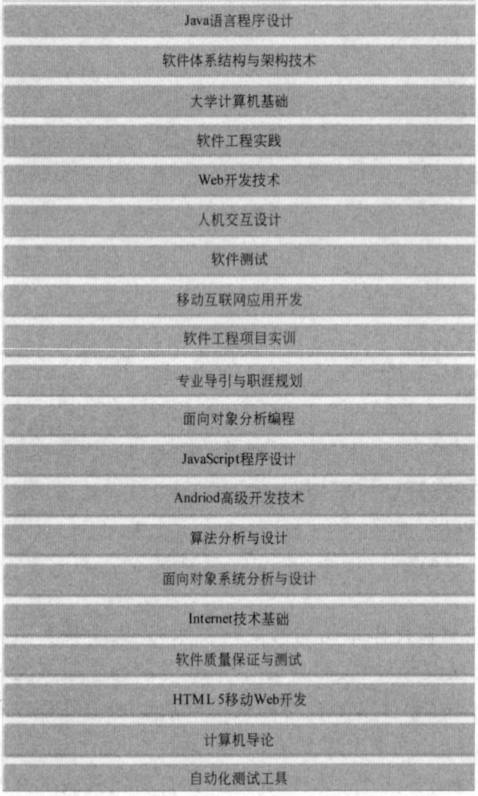

图 6.9　ScrollView 运行结果图 1 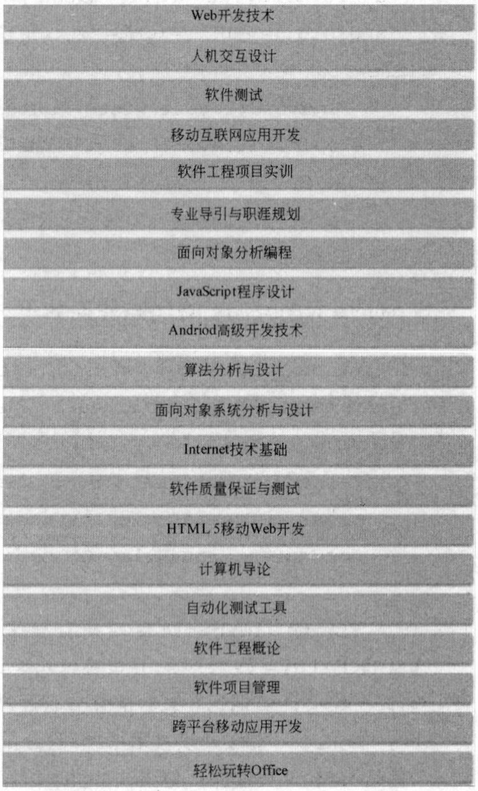　　　　图 6.10　ScrollView 运行结果图 2

6.3　按钮类控件

6.3.1　Button

Button 表示普通按钮，Button 控件继承自 TextView 类，用户可以对 Button 控件进行按下或单击等操作。对 Button 按钮的使用，主要是为 Button 控件设置 View.OnClickListener 监听器，并在监听的实现代码中开发按钮被按下事件的处理代码。

Button 控件除了可以在按钮上显示字符串外，还可以通过修改背景来显示图片等 Drawable 资源。

在Android中,可以应用两种方法向屏幕中添加按钮,一种是在Java文件中通过new关键字来创建,另一种是通过在XML布局文件中使用<Button>标记添加。后一种方法是比较常用的方法,添加普通按钮的基本格式如下。

```
<Button
        android:id="@+id/btn_login"
        android:layout_width="match_parent"
        android:layout_height="wrap_content"
        android:layout_marginLeft="50dp"
        android:layout_marginRight="50dp"
        android:layout_marginTop="50dp"
        android:textColor="#FFFFFF"
        android:textSize="17sp"
        android:text="@string/login"
        />
```

在屏幕中添加按钮后,按钮被单击是没有反应的,如果想让按钮发挥特有的用途,还需要为按钮添加事件处理的代码。Android 提供两种为按钮添加单击事件监听器的方法,一种是在Java代码中完成。例如,在Activity的onCreate()方法中完成,代码如下。

```
import android.view.View.OnClickListener;
import android.widget.Button;
Button btn1=(Button)findViewById(R.id.btn_login);
btn1.setOnClickListener(new OnClickListener(){
@Override
public void onClick(View arg0){
        //编写要执行的动作代码
            }
});
```

另一种是在Activity中编写包含View类型参数的方法,并且将要触发的动作代码放在该方法中,然后在布局文件中通过android:onClick属性指定对应的方法名实现。例如,在Activity中编写一个loginClick()方法,关键代码如下。

```
public void loginClick(View arg0){
        //编写要执行的动作代码
    }
```

这样就可以在布局文件中通过android:onClick="loginClick"为按钮添加单击事件监听器。

6.3.2 ImageButton

ImageButton表示图片按钮,ImageButton控件继承自ImageView类。在使用上,ImageButton控件与Button控件的不同之处在于ImageButton控件没有text属性,所以ImageButton控件不能显示文本。在ImageButton控件中设置按钮显示的图片,可以通过 setImageResource(int)方法来设置,也可以通过 android:src 属性来设置。

ImageButton 控件与 Button 控件默认的背景色是相同的,当按钮处于不同状态时,背景色也会随之变化。使用 ImageButton 控件时,一般将背景色设置为其他图片或直接设置为透明的,另外,使用该控件时需要为按钮控件指定不同状态下显示的图片,否则用户将无法区别是否按下了按钮。可以通过编写 XML 文件实现按钮在不同状态下显示不同图片的功能。

下面通过一个案例学习 ImageButton 控件的应用。

【例 6-4】 ImageButton 控件的应用。

本案例共创建三个 ImageButton 控件,第一个实现单击能切换图片的功能,第二个单击更换一张图片,第三个显示打电话图片。

实现步骤如下。

新建一个模块,名字为 ch06_4,打开项目文件夹中 res\layout 目录下的 activity_main.xml 文件,将其中的代码替换为以下代码。

```xml
<?xml version="1.0" encoding="utf-8"?>
<LinearLayout xmlns:android="http://schemas.android.com/apk/res/android"
    android:layout_width="fill_parent"
    android:layout_height="fill_parent"
    android:background="#FFEFEFEF"
    android:orientation="vertical"
    android:padding="10px"
    android:id="@+id/linear">
    <TextView
        android:layout_width="wrap_content"
        android:layout_height="wrap_content"
        android:layout_gravity="center_horizontal"
        android:text="selector 设置 ImageButton 图片"
        android:textSize="42px"
        android:textColor="#FF000000" />
    <ImageButton
        android:id="@+id/imgbut1"
        android:layout_width="wrap_content"
        android:layout_height="wrap_content"
        android:layout_gravity="center_horizontal"
        android:layout_margin="10px"
        android:background="@drawable/buttpic" />
    <TextView
        android:layout_width="wrap_content"
        android:layout_height="wrap_content"
        android:layout_gravity="center_horizontal"
        android:text="不同事件更改图片"
```

```
        android:textSize="42px"
        android:textColor="#FF000000" />
    <ImageButton
        android:id="@+id/imgbut2"
        android:layout_width="wrap_content"
        android:layout_height="wrap_content"
        android:layout_gravity="center_horizontal"
        android:layout_margin="10px"
        android:background="@drawable/butterfly" />
    <TextView
        android:layout_width="wrap_content"
        android:layout_height="wrap_content"
        android:layout_gravity="center_horizontal"
        android:text="调用系统默认图片"
        android:textSize="42px"
        android:textColor="#FF000000" />
    <ImageButton
        android:id="@+id/imgbut3"
        android:layout_width="wrap_content"
        android:layout_height="wrap_content"
        android:layout_gravity="center_horizontal"
        android:layout_margin="10px"
        android:background="@android:drawable/stat_sys_phone_call" />
</LinearLayout>
```

第一个图片按钮的切换图片功能需要使用<selector 属性设置</selector>来实现，此标签内容写在 buttpic.xml 文件中，代码如下。

```
<selector xmlns:android="http://schemas.android.com/apk/res/android">
<item android:state_pressed="false" android:drawable="@drawable/flower"/>
<item android:state_pressed="true" android:drawable="@drawable/grass"/>
</selector>
```

ImageButton 等控件的布局结构如图 6.11 所示。

图 6.11　ImageButton 等控件布局结构图

图 6.12 中(a)图所示为程序运行的初始状态,单击第一个 ImageButton 按钮时,切换为(b)图显示的状态。

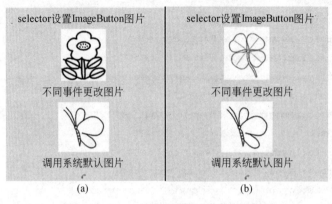

图 6.12　ImageButton 等控件运行结果图

6.3.3　ToggleButton

ToggleButton 控件继承于 Button,状态只能是选中和未选中,并且需要为不同的状态设置不同的显示文本。除了继承自父类的一些属性和方法外,ToggleButton 还具有一些自己的 XML 属性,如下所述。

android:checked:设置按钮是否被选中。

android:textOff:设置当该按钮没有被选中时显示的文本。

android:textOn:设置当该按钮被选中时显示的文本。

6.3.4　CheckBox

在默认情况下,复选框显示为一个方块图标,并且该图标旁边放置一些说明性文字。复选框能够进行多选设置,每一个复选框都提供"选中"和"不选中"两种状态。在 Android 中,复选框使用 CheckBox 表示,而 CheckBox 类又是 Button 的子类,所以复选框可以直接使用 Button 支持的各种属性。

在 Android 中,可以使用两种方法向屏幕中添加复选框,一种是通过在 XML 布局文件中使用<CheckBox>标记添加,另一种是在 Java 文件中通过 new 关键字创建,推荐使用第一种方法。在 XML 布局文件中添加复选框的格式如下。

```
<CheckBox
android:layout_height="wrap_content"
android:id="@+id/checkBox1"
android:button="@drawable/checkbox">
</CheckBox>
```

由于复选框可以选中多项,所以为了确定用户是否选中了某一项,还需要为每个选项添加事件监听器。例如,给 CheckBox 设置事件监听的代码如下。

```
CheckBox  checkBox1 = (CheckBox)findViewById(R.id.checkBox1);
//给 CheckBox 设置事件监听
checkBox1.setOnCheckedChangeListener(new CompoundButton.
OnCheckedChangeListener(){
            @Override
            public void onCheckedChanged(CompoundButton buttonView,
                    boolean isChecked){
                //TODO Auto-generated method stub
                if(isChecked){
                    checkBox1.setText("选中");
                }else{
                    checkBox1.setText("取消选中");
                }
            }
        });
```

6.3.5　RadioButton

RadioButton 表示单选按钮,是 Button 的子类,所以单选按钮继承了 Button 的各种属性,可以直接使用。默认情况下,单选按钮显示为一个圆形图标,并且该图标旁边放置一些说明性文字。而在程序中,一般将多个单选按钮放置在按钮组中,使这些单选按钮能够实现互斥,当用户选中某个单选按钮后,按钮组中的其他按钮将自动取消选中状态。

RadioGroup 是单选组合框,用于将 RadioButton 框起来;在没有 RadioGroup 的情况下,RadioButton 可以全部都选中;当多个 RadioButton 被 RadioGroup 包含时,RadioButton 只可以选择一个。

```
<RadioGroup
    android:id="@+id/radiogroup1"
    android:layout_width="wrap_content"
    android:layout_height="wrap_content"
    android:orientation="vertical"
    android:layout_x="3px"
>
    <RadioButton
        android:id="@+id/radiobutton1"
        android:layout_width="wrap_content"
        android:layout_height="wrap_content"
        android:text="@string/radiobutton1"
    />
    <RadioButton
        android:id="@+id/radiobutton2"
        android:layout_width="wrap_content"
        android:layout_height="wrap_content"
        android:text="@string/radiobutton2"
```

```
/>
</RadioGroup>
```

RadioButton 控件的 android:checked 属性用于指定选中状态,当属性值为 true 时,表示选中,当属性值为 false 时,表示没有被选中,默认值为 false。

用 setOnCheckedChangeListener 来对单选按钮进行监听。举例如下。

```
RadioGroup radiogroup=(RadioGroup)findViewById(R.id.radiogroup1);
  RadioButton   radio1=(RadioButton)findViewById(R.id.radiobutton1);
  RadioButton   radio2=(RadioButton)findViewById(R.id.radiobutton2);
  radiogroup.setOnCheckedChangeListener(new RadioGroup.
  OnCheckedChangeListener(){

      @Override
      public void onCheckedChanged(RadioGroup group, int checkedId){
        //TODO Auto-generated method stub
        if(checkedId==radio2.getId())
        {
    Toast.makeText(AddShopActivity.this,"回答正确",
            Toast.LENGTH_SHORT).show();
          }else
    {
    Toast.makeText(AddShopActivity.this,"回答错误",
            Toast.LENGTH_SHORT).show();

        }
      }
    });
}
```

【例 6-5】 按钮类控件的应用。

本案例介绍了 Button 控件的综合应用,涉及的 Button 类控件有 Button、ImageButton、ToggleButton、CheckBox 和 RadioButton。本案例实现了 ImageButton 控件的显示图片功能,ToggleButton 按钮的状态切换功能,CheckBox 的多选和 RadioButton 的单选功能,最后对 Button 按钮做了事件处理,获取并输出了 CheckBox 和 RadioButton 的选中信息,具体的布局文件内容如下。

```
<?xml version="1.0" encoding="utf-8"?>
<LinearLayout xmlns:android="http://schemas.android.com/apk/res/android"
    android:layout_width="match_parent"
    android:layout_height="match_parent"
    android:orientation="vertical">
<Button
        android:layout_width="match_parent"
        android:layout_height="wrap_content"
```

```xml
        android:onClick="buttonClick"
        android:text="Common Button" />
<ImageButton
        android:layout_width="wrap_content"
        android:layout_height="wrap_content"
        android:src="@drawable/button_image" />
<Button
        android:layout_width="wrap_content"
        android:layout_height="wrap_content"
        android:background="@drawable/button_image"
        android:onClick="buttonClick" />
<ToggleButton
        android:layout_width="wrap_content"
        android:layout_height="wrap_content"
        android:checked="true"
        android:textOff="关的状态"
        android:textOn="开的状态" />
<TextView
        android:layout_width="match_parent"
        android:layout_height="wrap_content"
        android:text="请选择您所在的城市:" />
<RadioGroup
        android:id="@+id/radioContainer"
        android:layout_width="match_parent"
        android:layout_height="wrap_content"
        android:orientation="horizontal">
<RadioButton
            android:layout_width="wrap_content"
            android:layout_height="wrap_content"
            android:text="北京" />
<RadioButton
            android:layout_width="wrap_content"
            android:layout_height="wrap_content"
            android:text="上海" />
<RadioButton
            android:layout_width="wrap_content"
            android:layout_height="wrap_content"
            android:text="大连" />
</RadioGroup>
<TextView
        android:layout_width="match_parent"
        android:layout_height="wrap_content"
        android:text="请选择您到过的城市:" />
<LinearLayout
```

```xml
        android:id="@+id/checkContainer"
        android:layout_width="match_parent"
        android:layout_height="wrap_content"
        android:orientation="horizontal">
    <CheckBox
            android:layout_width="wrap_content"
            android:layout_height="wrap_content"
            android:text="北京" />
    <CheckBox
            android:layout_width="wrap_content"
            android:layout_height="wrap_content"
            android:text="上海" />
    <CheckBox
            android:layout_width="wrap_content"
            android:layout_height="wrap_content"
            android:text="大连" />
</LinearLayout>
<Button
        android:layout_width="wrap_content"
        android:layout_height="wrap_content"
        android:onClick="btnDisplaySelected"
        android:text="获取选中" />
<Switch
        android:id="@+id/switch1"
        android:layout_width="wrap_content"
        android:layout_height="wrap_content"
        android:text="Switch" />
</LinearLayout>
```

MainActivity.java 中的主要代码如下。

```java
public class MainActivity extends AppCompatActivity {
    private RadioGroup radioContainer;
    private LinearLayout checkContainer;
    @Override
    protected void onCreate(Bundle savedInstanceState) {
        super.onCreate(savedInstanceState);
        setContentView(R.layout.activity_main);
        radioContainer= (RadioGroup)(findViewById(R.id.radioContainer));
        checkContainer= (LinearLayout)(findViewById(R.id.checkContainer));
    }
    public void buttonClick(View view){
        Log.i("Button", "Button Clicked");
    }
    public void btnDisplaySelected(View view){
```

```java
        String radioText="";
        for(int i=0;i<radioContainer.getChildCount();i++)
        {
            View c=radioContainer.getChildAt(i);
            if(c instanceof RadioButton){
                RadioButton radio=(RadioButton)c;
                if(radio.isChecked()){
                    radioText=radio.getText().toString();
                    break;
                }
            }
        }
        Log.i("Radio", "Radio Selected is "+radioText);
        String checkText="";
        for(int i=0;i<checkContainer.getChildCount();i++){
            View c=checkContainer.getChildAt(i);
            if(c instanceof CheckBox){
                CheckBox check=(CheckBox)c;
                if(check.isChecked()){
                    checkText +=check.getText().toString()+";";
                }
            }
        }
        Log.i("Check", "Check Selected is "+checkText);
    }
```

布局结构如图6.13所示。

图6.13 Button等控件布局结构图

运行结果如图 6.14 所示，(a) 图是初始状态，(b) 图是对屏幕内容进行了选择后的效果。

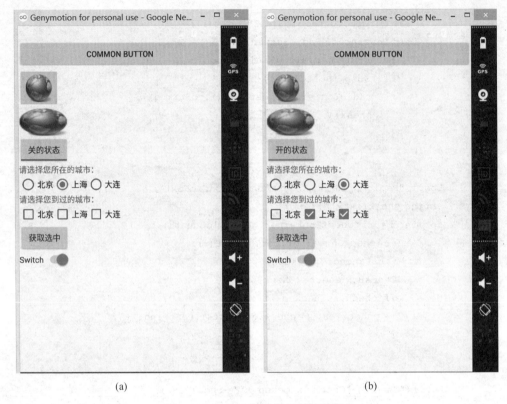

图 6.14　Button 等控件运行结果图 1

当单击"获取选中"按钮时，调用程序中的事件处理方法，在 logcat 窗口显示程序的运行结果，即多选和单选按钮的选中状态，如图 6.15 所示。

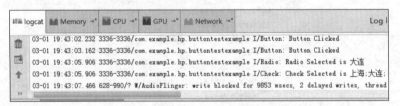

图 6.15　ImageButton 等控件运行结果图 2

6.4　日期和时间类控件

6.4.1　DatePicker

DatePicker 继承自 FrameLayout 类，日期选择控件的主要功能是向用户提供包含年、月、日的日期数据，并允许用户对其修改。如果要捕获用户修改日期选择控件中的数据事件，需要为 DatePicker 添加 OnDateChangedListener 监听器。代码如下。

```
<DatePicker android:id="@+id/datePicker"
android:layout_width="wrap_content"
android:layout_height="wrap_content"
android:layout_gravity="center_horizontal"/>
DatePicker datePicker=(DatePicker)findViewById(R.id.datePicker);
Calendar calendar=Calendar.getInstance();
int year=calendar.get(Calendar.YEAR);
int monthOfYear=calendar.get(Calendar.MONTH);
int dayOfMonth=calendar.get(Calendar.DAY_OF_MONTH);
datePicker.init(year, monthOfYear, dayOfMonth, new OnDateChangedListener(){
public void onDateChanged(DatePicker view, int year,
int monthOfYear, int dayOfMonth){
dateEt.setText("您选择的日期是："+year+"年"+(monthOfYear+1)+"月"+dayOfMonth
+"日。");}
});
```

具体的应用请看下面的例子。

【例 6-6】 DatePicker 控件的应用。

本案例介绍 DatePicker 控件的应用。首先在布局文件中添加一个按钮控件，单击按钮触发事件，处理生成 DatePickerDialog 的对象，并创建 OnDateSetListener 的 onDateSet()方法，从而实现设置日期的功能。

实现步骤如下。

新建一个模块，名字为 ch06_6，打开项目文件夹中 res\layout 目录下的 activity_main.xml 文件，将其中的代码替换为以下代码。

XML 文件中的主要代码如下所示。

```
<?xml version="1.0" encoding="utf-8"?>
<LinearLayout xmlns:android="http://schemas.android.com/apk/res/android"
    android:layout_width="match_parent"
    android:layout_height="match_parent"
    android:background="#FFFFFFFF"
    android:orientation="vertical">
<TextView
        android:id="@+id/showDate"
        android:layout_width="wrap_content"
        android:layout_height="wrap_content"
        android:layout_weight="0"
        android:textColor="#FF000000" />
<Button
        android:id="@+id/setDate"
        android:layout_width="wrap_content"
```

```
            android:layout_height="wrap_content"
            android:text="设置日期" />
</LinearLayout>
```

MainActivity.java 中的主要代码如下。

```java
public class MainActivity extends AppCompatActivity {
    private TextView showDate;
    private Button setDate;
    private int year;
    private int month;
    private int day;
    @Override
    protected void onCreate(Bundle savedInstanceState){
        super.onCreate(savedInstanceState);
        setContentView(R.layout.activity_main);
        showDate= (TextView)this.findViewById(R.id.showDate);   //获取用来显示当前日
                                                                //期的 TextView 组件
        setDate= (Button)this.findViewById(R.id.setDate);   //获取 Button 组件
        //初始化 Calendar 日历对象
      /* Calendar myCalendar=Calendar.getInstance(Locale.CHINA);
        Date myDate=new Date();                     //获取当前日期 Date 对象
        myCalendar.setTime(myDate);                 //为 Calendar 对象设置时间为当前日期
        year=myCalendar.get(Calendar.YEAR);         //获取 Calendar 对象中的年
        month=myCalendar.get(Calendar.MONTH);       //获取 Calendar 对象中的月,
                                                    //0 表示 1 月,1 表示 2 月...
        day=myCalendar.get(Calendar.DAY_OF_MONTH);  //获取这个月的第几天
        showDate.setText(year+"-"+(month+1)+"-"+day);  //修改 TextView 显示的信息
                                                       //为当前的年月日
      */
        showDate.setText(new SimpleDateFormat("yyyy-MM-dd").format(new Date()));
        setDate.setOnClickListener(new OnClickListener(){   //"设置日期"按钮的单
                                                            //击事件

            @Override
            public void onClick(View v){
                //TODO Auto-generated method stub
                /* 创建 DatePickerDialog 对象
                   构造函数原型:
public DatePickerDialog (Context context, DatePickerDialog.OnDateSetListener
callBack, int year, int monthOfYear, int dayOfMonth)
参数含义依次为 context:组件运行 Activity,DatePickerDialog.OnDateSetListener:选
择日期事件
year:当前组件上显示的年,monthOfYear:当前组件上显示的月,dayOfMonth:当前组件上显
```

示的日 */
```
                DatePickerDialog dpd=new DatePickerDialog
                    (MainActivity.this,
                        new OnDateSetListener(){
                            /*
                             * view:该事件关联的组件
                             * myyear:当前选择的年
                             * monthOfYear:当前选择的月
                             * dayOfMonth:当前选择的日
                             */
                        @Override
                    public void onDateSet(
                DatePicker view, int myyear, int monthOfYear, int dayOfMonth){
                    //TODO Auto-generated method stub
                    //在DatePickerDialog组件上设置日期后,同时修改TextView上的信息
                    showDate.setText(myyear+"-"+(monthOfYear+1)+"-"+dayOfMonth);
                    //修改year,month,day变量值,以便在依次单击按钮时DatePickerDialog
                    //显示上一次修改后的值
                    year=myyear;
                    month=monthOfYear;
                    day=dayOfMonth;
                }
            }, year, month, day);
            dpd.show();                                         //显示DatePickerDialog组件
                }
            });
        }
    }
```

DatePicker案例的布局结构如图6.16所示,由此结构图可见,初始的布局中并没有DatePicker控件,该控件是在程序运行中添加进来的。

运行结果如图6.17所示。

图6.16 DatePicker的布局结构图

图6.17 DatePicker的运行结果图1

单击设置日期按钮,出现如图6.18所示的控件的显示界面,就可以进行日期的设置了。

图 6.18 DatePicker 的运行结果图 2

6.4.2 TimePicker

TimePicker 与 DatePicker 一样,也继承自 FrameLayout 类,TimePicker 控件是向用户显示一天中的时间(可以为 24 小时制,也可以为 AM/PM 制),并允许用户进行选择。如果要捕获用户修改时间数据的事件,便需要为 TimePicker 添加 OnTimeChangedListener 监听器。

具体的应用请看下面的例子。

【例 6-7】 TimePicker 控件的应用。

本案例介绍 TimePicker 控件的应用,首先在布局文件中添加一个按钮控件,按钮显示"设置时间",单击按钮触发事件,处理生成 TimePickerDialog 的对象,并覆盖 OnTimeChangedListener 的 onTimeChanged() 方法,从而实现设置时间的功能。

实现步骤如下。

新建一个模块,名字为 ch06_7,打开项目文件夹中 res\layout 目录下的 activity_main.xml 文件,将其中的代码替换为以下代码。

XML 文件中的主要代码如下。

```
<?xml version="1.0" encoding="utf-8"?>
```

```xml
<LinearLayout xmlns:android="http://schemas.android.com/apk/res/android"
    android:layout_width="match_parent"
    android:layout_height="match_parent"
    android:background="#FFFFFFFF"
    android:orientation="vertical">
<LinearLayout
        android:layout_width="match_parent"
        android:layout_height="0px"
        android:layout_weight="1"
        android:orientation="vertical">
<TextView
          android:id="@+id/showTime"
          android:layout_width="wrap_content"
          android:layout_height="wrap_content"
          android:textColor="#FF000000" />
</LinearLayout>
<Button
        android:id="@+id/setTime"
        android:layout_width="wrap_content"
        android:layout_height="wrap_content"
        android:onClick="set"
        android:text="设置日期" />
</LinearLayout>
```

MainActivity.java 中的主要代码如下。

```java
public class TimePickerDialogExample extends Activity {
   private TextView showTime;
   private Button setTime;
   private int year;
   private int month;
   private int day;
   private int hour;
   private int minus;
   public void set(View b){
   }
   /** Called when the activity is first created. */
   @Override
   public void onCreate(Bundle savedInstanceState){
       super.onCreate(savedInstanceState);
       setContentView(R.layout.main);          //加载 activity_main.xml 布局文件
       showTime= (TextView)this.findViewById(R.id.showTime);
                                  //获取用来显示当前时间的 TextView 组件
       setTime= (Button)this.findViewById(R.id.setTime);
                                  //获取设置时间 Button 组件
       Calendar myCalendar=Calendar.getInstance(Locale.CHINA);
```

```java
                                        //初始化Calendar日历对象
        Date myDate=new Date();         //获取当前日期Date对象
        myCalendar.setTime(myDate);     //为Calendar对象设置时间为当前日期
        year=myCalendar.get(Calendar.YEAR);    //获取Calendar对象中的年
        month=myCalendar.get(Calendar.MONTH);
                                //获取Calendar对象中的月,0表示1月,1表示2月...
        day=myCalendar.get(Calendar.DAY_OF_MONTH);
                                        //获取这个月的第几天
        hour=myCalendar.get(Calendar.HOUR_OF_DAY);
                                        //获取小时信息
        minus=myCalendar.get(Calendar.MINUTE);   //获取分钟信息
        showTime.setText(year+"-"+(month+1)+"-"+day+" "+hour+":"+minus);
                                        //设置TextView组件上显示的日期信息
        setTime.setOnClickListener(new OnClickListener(){
            @Override
            public void onClick(View v){
                TimePickerDialog tpd=new TimePickerDialog(TimePickerDialogExample.this,
                    new OnTimeSetListener(){
                        @Override
        public void onTimeSet(TimePicker view, int hourOfDay, int minute)
    {   showTime.setText(year+"-"+month+"-"+day+"-"+hourOfDay+"-"+minute);
                hour=hourOfDay;
                minus=minute;
                    }
                }, hour, minus, true);
                tpd.show();
            }
        });
        setTime.setOnClickListener(new OnClickListener(){
                                        //"设置日期"按钮的单击事件
            @Override
            public void onClick(View v){
                //TODO Auto-generated method stub
                //创建TimePickerDialog对象
                //构造函数原型:TimePickerDialog(Context context, TimePickerDialog.
                    OnTimeSetListener callBack, int hourOfDay, int minute, boolean
                    is24HourView)
                //参数含义依次为context:组件运行Activity,TimePickerDialog.
                    OnTimeSetListener:选择时间事件
                //hourOfDay:当前组件上显示小时,minute:当前组件上显示分钟,
                    is24HourView:是否24小时方式显示,或者AM/PM方式显示
                TimePickerDialog tpd=new TimePickerDialog(TimePickerDialogExample.
                    this,new TimePickerDialog.OnTimeSetListener(){
                        @Override
                        public void onTimeSet(TimePicker view, int hourOfDay,
                                    int myminute){
                            //TODO Auto-generated method stub
```

```
            showTime.setText(year+"-"+(month+1)+"-"+day+" "+hourOfDay
            +":"+myminute);
            hour=hourOfDay;
            minus=myminute;
            }
        },hour,minus,false);
        tpd.show();
        }
    });
    }
}
```

TimePicker 案例布局结构如图 6.19 所示,此布局中包括一个 TextView,用来显示时间,还有一个 Button,单击则显示 TimePicker 的界面,对时间进行设置。

图 6.19　TimePicker 的布局结构图

运行结果如图 6.20 所示,当单击图 6.20 中的"设置日期"按钮时,弹出如图 6.21 所示的界面。

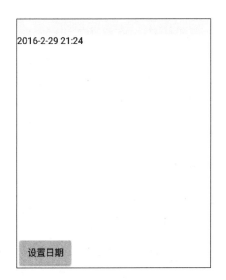

图 6.20　TimePicker 的运行结果图 1　　　　图 6.21　TimePicker 的运行结果图 2

6.4.3 DigitalClock

DigitalClock 控件显示时间，能够实时更新。在 Android 中，时钟控件包括 AnalogClock 和 DigitalClock，它们都负责显示时钟，不同的是 AnalogClock 控件显示模拟时钟，且只显示时针和分针，而 DigitalClock 显示数字时钟，可精确到秒，DigitalClock 在布局文件中显示的代码如下。

```
<DigitalClock
    android:layout_width="wrap_content"
    android:layout_height="wrap_content"
    android1:layout_centerVertical="true"
    android:layout_alignParentLeft="true"
    android:textSize="30sp"
    android:layout_gravity="center_horizontal"/>
```

6.4.4 Chronometer

Chronometer 集成自 TextView，里面有个 Handler，负责定时更新 UI。其计时原理很简单：通过 setBase(long t) 方法设置好 baseTime 之后，当 start() 时，每隔 1 秒，用当前 SystemClock.elapsedRealtime() 减 baseTime，得到的逝去时间显示在 TextView 中。

下面举个例子，这里除了 start、stop 功能，还利用 setBase() 添加了 pause 功能。

```
@Override
public void onClick(View v){
    switch(v.getId()){
        case R.id.startButton:
            //保证什么时候开始，计时的时间都是从 0 开始计时
            recordChronometer.start();
            recordChronometer.setBase(SystemClock.elapsedRealtime());
            break;
        case R.id.pauseButton:
            //实现了计时的暂停功能，如果不是处于暂停状态，则停止计时
            //如果处于暂停状态，则通过计算，从暂停的那个时间开始计时
            if(!isPause){
                recordChronometer.stop();
                isPause=true;
                pauseButton.setText("继续计时");
            } else {
                Double temp=Double.parseDouble(recordChronometer.getText()
                    .toString().split(":")[1] * 1000;
                recordChronometer
                    .setBase((long)(SystemClock.elapsedRealtime()-temp));
                recordChronometer.start();
                pauseText();
```

```
            }
            break;
        case R.id.stopButton:
            recordChronometer.start();
            recordChronometer.setBase(SystemClock.elapsedRealtime());
            break;
        case R.id.resetButton:
            //停止后,使时间归零
            recordChronometer.stop();
            recordChronometer.setBase(SystemClock.elapsedRealtime());
            pauseText();
            timer.setText("获得的时间为:"+recordChronometer.getText().
            toString());
            break;
        default:
            break;
    }
}
//设置处于暂停状态时,pause按钮的文字显示
public void pauseText(){
    if(isPause){
        pauseButton.setText("暂停计时");
        isPause=false;
    }
}
```

【例6-8】 Chronometer、AnalogClock 和 DigitalClock 控件的应用。

本案例介绍 Chronometer、AnalogClock 和 DigitalClock 控件的应用,通过此案例掌握三种控件的创建方式和在界面显示的效果。具体应用请看下面的例子。

```
<?xml version="1.0" encoding="utf-8"?>
<LinearLayout xmlns:android="http://schemas.android.com/apk/res/android"
    android:layout_width="match_parent"
    android:layout_height="match_parent"
    android:orientation="vertical">
    <Chronometer
        android:id="@+id/chronometer1"
        android:layout_width="240dp"
        android:layout_height="wrap_content"
        android:text="Chronometer" />
    <AnalogClock
        android:id="@+id/analogClock1"
        android:layout_width="wrap_content"
        android:layout_height="wrap_content" />
    <DigitalClock
        android:id="@+id/digitalClock1"
        android:layout_width="wrap_content"
```

```xml
            android:layout_height="wrap_content"
            android:text="DigitalClock" />
    <Button
            android:layout_width="wrap_content"
            android:layout_height="wrap_content"
            android:id="@+id/buttonstart"
            android:text="启动" />
    <Button
            android:layout_width="wrap_content"
            android:layout_height="wrap_content"
            android:id="@+id/buttonstop"
            android:text="停止" />
</LinearLayout>
```

程序的布局结构如图 6.22 所示。

图 6.22 案例的布局结构

MainActivity 中的代码如下。

```java
import android.os.Bundle;
import android.os.SystemClock;
import android.support.v7.app.AppCompatActivity;
import android.view.View;
import android.widget.Button;
import android.widget.Chronometer;
public class MainActivity extends AppCompatActivity
        implements View.OnClickListener {
    Chronometer recordChronometer;
    Button startButton, stopButton;
    boolean isPause=false;                    //用于判断是否为暂停状态
    long recordingTime=0;
    @Override
    protected void onCreate(Bundle savedInstanceState){
        super.onCreate(savedInstanceState);
        setContentView(R.layout.activity_main);
        recordChronometer= (Chronometer)findViewById(R.id.chronometer1);
        startButton= (Button)findViewById(R.id.buttonstart);
        stopButton= (Button)findViewById(R.id.buttonstop);
        startButton.setOnClickListener(this);
```

```
        stopButton.setOnClickListener(this);
    }
    @Override
    public void onClick(View v){
        switch(v.getId()){
            case R.id.buttonstart:
                //保证什么时候开始,计时的时间都是从0开始计时
                recordChronometer.start();
                recordChronometer.setBase(SystemClock.elapsedRealtime());
                break;
            case R.id.buttonstop:
                //停止后,使时间归零
                recordChronometer.stop();
                recordChronometer.setBase(SystemClock.elapsedRealtime());
                break;
            default:
                break;
        }
    }
}
```

程序运行结果如图 6.23 所示。

图 6.23　案例的布局结构

6.5 进度条控件 ProgressBar

进度条是 UI 界面中一种非常实用的组件，用于向用户显示某个比较耗时操作完成的百分比。因此，进度条可以动态地显示进度，避免长时间地执行某个耗时操作，让用户感觉程序失去了响应，从而提高界面的友好性。

Android 中的进度条使用 ProgressBar 表示，用于向用户显示某个耗时操作完成的百分比。

Android 支持几种风格的进度条，通过 style 属性可以为 Progress 指定风格。该属性支持如下几个属性值。

- ＃@android:style/Widget.ProgressBar.Horizontal：水平进度条。
- ＃@android:style/Widget.ProgressBar.Inverse：普通大小的环形进度条。
- ＃@android:style/Widget.ProgressBar.Large：大环形进度条。
- ＃@android:style/Widget.ProgressBar.Large.Inverse：大环形进度条。
- ＃@android:style/Widget.ProgressBar.Small：小环形进度条。
- ＃@android:style/Widget.ProgressBar.Small.Inverse：小环形进度条。

除此之外，ProgressBar 还支持以下所示的常用 XML 属性值。

android:max：设置该进度条的最大值。

android:progress：设置该进度条已完成的进度值。

android:progressDrawable：设置该进度条轨道的绘制形式。

android:progressBarStyle：设置该进度条的默认进度样式。

android:progressBarStyleLarge：设置大进度条样式。

android:progressBarStyleSmall：设置小进度条样式。

ProgressBar 使用格式如以下代码所示。

```
<ProgressBar
属性设置 />
```

使用进度条时，还需调用常用方法，常用方法如下。

getMax()：返回这个进度条范围的上限。

getProgress()：返回进度。

getSecondaryProgress()：返回次要进度。

incrementProgressBy(int diff)：指定增加的进度或减少（＋进度增加，－进度减少）。

isIndeterminate()：指示进度条是否在不确定模式下。

setIndeterminate(boolean indeterminate)：设置不确定模式。

setVisibility(int v)：设置该进度条是否可视。

6.6 滑动条 SeekBar

拖动条类似进度条，不同的是用户可以控制。比如，在应用程序中，用户可以控制音效，这就可以使用拖动条来实现。

在 Android 中，如果想在屏幕中添加拖动条，可在 XML 布局文件中通过
＜SeekBar＞标记添加，格式如下。

```
<SeekBar android:id="@+id/seekbar"
    android:layout_width="match_parent"
    android:layout_height="wrap_content"/>
```

改变拖动滑块的外观，可以使用 android：thumb 属性实现，该属性的值为一个
Drawble 对象，该 Drawble 对象将作为自定义滑块。

由于拖动条控件允许用户改变拖动滑块的外观，因此需要对其进行事件监听，这就需
要实现 SeekBar.OnSeekBarChangeListener 接口。在 SeekBar 中需要监听 3 个事件，分
别是数值的改变（onProgressChanged）、开始拖动（onStartTrackingTouch）、停止拖动
（onStopTrackingTouch）。在 onProgressChanged 中可以得到当前数值的大小。

6.7　星级控件 RatingBar

在 Android 中，星级评分条使用 RatingBar 表示，和拖动条类似，星级评分条允许用
户通过拖动来改变进度，不同的是，星级评分条通过星形表示进度，通常使用星级评分条
表示对某一事物的支持度或对某种服务的满意程度等。

在 Android 中，如果想在屏幕中添加星级评分条，可在 XML 布局文件中通过
＜RatingBar＞标记添加，格式如下。

```
<RatingBar
    android:id="@+id/ratingbar"
    android:layout_width="wrap_content"
    android:layout_height="match_parent"
    android:numStars="5"
    android:stepSize="0.5"
    />
```

RatingBar 控件支持的 XML 属性如下。

android：numStars：用于指定该评分条星的数量。

android：stepSize：用于指定每次最少需要改变多少个星级，默认为 0.5。

android：isIndicator：用于指定该星级评分条是否允许用户改变，true 为不允许改变。

android：rating：用于指定该星级评分条默认的星级。

除此之外，星级评分条还提供了 3 个比较常用的方法。

getRating()方法：用于获取等级，表示被选中了几颗星。

getStepSize()方法：用于获取每次最少要改变多少个星级。

getProgress()方法：用于获取进度，获取到的进度值等于 getRating()方法的返回值
与 getStepSize()方法的返回值的乘积。

【例 6-9】　进度条控件的应用。

本案例介绍 ProgressBar 和 SeekBar 控件的应用，给进度条设置了样式，所以需要在

drawable 文件夹中添加 3 个样式 XML 文件,分别是 barcolor1.xml、barcolor2.xml、Barface.xml。布局文件中添加了 5 个进度条,2 个 ProgressBar 和 3 个 SeekBar。具体应用请看下面的例子。

案例组成如下。

barcolor1.xml 文件内容如下。

```
<?xml version="1.0" encoding="UTF-8"?>
<layer-list xmlns:android="http://schemas.android.com/apk/res/android">
<item
    android:id="@android:id/background"
    android:drawable="@drawable/bg" />
<item
    android:id="@android:id/secondaryProgress"
    android:drawable="@drawable/secondary" />
<item
    android:id="@android:id/progress"
    android:drawable="@drawable/progress" />

</layer-list>
```

barcolor2.xml 文件内容如下。

```
<?xml version="1.0" encoding="UTF-8"?>
<layer-list xmlns:android="http://schemas.android.com/apk/res/android">
<item android:id="@android:id/background">
<shape>
<corners android:radius="10dip" />
<gradient
            android:angle="270"
            android:centerColor="#FF880000"
            android:centerY="0.75"
            android:endColor="#FF110000"
            android:startColor="#FFFF0000" />
</shape>
</item>
<item android:id="@android:id/secondaryProgress">
<clip>
<shape>
<corners android:radius="10dp" />
<gradient
            android:angle="270"
            android:centerColor="#FF00FF00"
            android:centerY="0.75"
            android:endColor="#FF00FF00"
            android:startColor="#FF00FF00" />
```

```
        </shape>
    </clip>
</item>
<item android:id="@android:id/progress">
<clip>
<shape>
<corners android:radius="10dp" />
<gradient
                android:angle="270"
                android:centerColor="#ffffb600"
                android:centerY="0.75"
                android:endColor="#ffffcb00"
                android:startColor="#ffffd300" />
        </shape>
    </clip>
</item>

</layer-list>
```

Barface.xml 文件内容如下。

```
<?xml version="1.0" encoding="UTF-8"?>
<layer-list xmlns:android="http://schemas.android.com/apk/res/android">
<item
        android:id="@android:id/background"
        android:drawable="@drawable/bg" />
<item
        android:id="@android:id/progress"
        android:drawable="@drawable/face" />
</layer-list>
```

布局文件内容如下。

```
<?xml version="1.0" encoding="utf-8"?>
<LinearLayout xmlns:android="http://schemas.android.com/apk/res/android"
    android:layout_width="match_parent"
    android:layout_height="match_parent"
    android:orientation="vertical">
<TextView
        android:layout_width="match_parent"
        android:layout_height="wrap_content"
        android:layout_marginTop="5dp"
        android:text="barcolor1" />
<ProgressBar
        android:id="@+id/progressBarHorizontal1"
        style="?android:attr/progressBarStyleHorizontal"
```

```xml
            android:layout_width="match_parent"
            android:layout_height="wrap_content"
            android:max="100"
            android:progress="30"
            android:progressDrawable="@drawable/barcolor1"
            android:secondaryProgress="60" />
    <TextView
            android:layout_width="match_parent"
            android:layout_height="wrap_content"
            android:layout_marginTop="5dp"
            android:text="barcolor2" />
    <ProgressBar
            android:id="@+id/progressBarHorizontal2"
            style="?android:attr/progressBarStyleHorizontal"
            android:layout_width="match_parent"
            android:layout_height="wrap_content"
            android:max="100"
            android:progress="30"
            android:progressDrawable="@drawable/barcolor2"
            android:secondaryProgress="60" />
    <TextView
            android:layout_width="match_parent"
            android:layout_height="wrap_content"
            android:layout_marginTop="5dp"
            android:text="barcolor1" />
    <SeekBar
            android:layout_width="match_parent"
            android:layout_height="wrap_content"
            android:max="100"
            android:progress="30"
            android:progressDrawable="@drawable/barcolor1" />
    <TextView
            android:layout_width="match_parent"
            android:layout_height="wrap_content"
            android:layout_marginTop="5dp"
            android:text="barcolor2" />
    <SeekBar
            android:layout_width="match_parent"
            android:layout_height="wrap_content"
            android:max="100"
            android:progress="30"
            android:progressDrawable="@drawable/barcolor2" />
    <TextView
            android:layout_width="match_parent"
```

```
        android:layout_height="wrap_content"
        android:layout_marginTop="5dp"
        android:text="Face拖动条" />
    <SeekBar
        android:layout_width="match_parent"
        android:layout_height="wrap_content"
        android:max="100"
        android:progress="30"
        android:progressDrawable="@drawable/barface" />
</LinearLayout>
```

程序的控件布局结构如图 6.24 所示。

运行结果如图 6.25 所示。

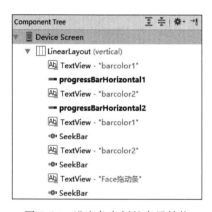

图 6.24　进度条案例的布局结构

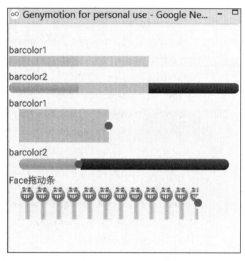

图 6.25　进度条案例的运行结果

【例 6-10】 进度条和 Clock 控件的综合应用。

本案例介绍 ProgressBar、SeekBar 和 RatingBar 控件以及 Chronometer、DigitalClock 和 AnalogClock 的综合应用,具体应用请看下面的例子。

实现步骤如下。

新建一个模块,名称为 ch06_10,打开项目文件夹中 res\layout 目录下的 activity_main.xml 文件,将其中的代码替换为以下代码。

```
<?xml version="1.0" encoding="utf-8"?>
<?xml version="1.0" encoding="utf-8"?>
<LinearLayout xmlns:android="http://schemas.android.com/apk/res/android"
    android:layout_width="match_parent"
    android:layout_height="match_parent"
    android:orientation="vertical"
    android:weightSum="1">
```

```xml
<ProgressBar
    style="?android:attr/progressBarStyleLarge"
    android:layout_width="match_parent"
    android:layout_height="wrap_content"
    android:id="@+id/progressBar"
    android:layout_weight="0.02" />
<ProgressBar
    style="?android:attr/progressBarStyleSmall"
    android:layout_width="match_parent"
    android:layout_height="wrap_content"
    android:id="@+id/progressBar2"
    />
<ProgressBar
    style="?android:attr/progressBarStyleHorizontal"
    android:layout_width="match_parent"
    android:layout_height="wrap_content"
    android:id="@+id/progressBar3"
    android:max="100"
    android:progress="25"
    android:layout_gravity="center_horizontal"
    android:layout_weight="0.02"
    />
<SeekBar
    android:layout_width="match_parent"
    android:layout_height="wrap_content"
    android:id="@+id/seekBar"
    android:max="100"
    android:progress="30"
    android:layout_weight="0.02"
    android:progressDrawable="@drawable/barface"
    />
<!--android:progressDrawable="@drawable/barface"-->
<RatingBar
    android:layout_width="wrap_content"
    android:layout_height="wrap_content"
    android:id="@+id/ratingBar2"
    android:layout_weight="0.02" />
<Switch
    android:layout_width="wrap_content"
    android:layout_height="wrap_content"
    android:text="New Switch"
    android:id="@+id/switch2"
    android:layout_weight="0.02" />
<RatingBar
```

```xml
    android:layout_width="wrap_content"
    android:layout_height="wrap_content"
    android:id="@+id/ratingBar"
    android:numStars="6"
    android:rating="2.5"
    android:layout_weight="0.02" />
<Chronometer
    android:layout_width="95dp"
    android:layout_height="wrap_content"
    android:id="@+id/chronometer"
    android:layout_weight="0.02"
    />
<TableLayout
    android:layout_width="match_parent"
    android:layout_height="wrap_content"
    android:layout_gravity="right"
    android:layout_weight="0.02"></TableLayout>
<AnalogClock
    android:layout_width="99dp"
    android:layout_height="99dp"
    android:id="@+id/analogClock" />
<LinearLayout
    android:orientation="horizontal"
    android:layout_width="match_parent"
    android:layout_height="64dp">
    <Button
        style="@style/buttonstyle"
        android:text="启动"
        android:id="@+id/button1"
        />
    <Button
        style="@style/buttonstyle"
        android:text="暂停"
        android:id="@+id/button2"
        />
    <Button
        style="@style/buttonstyle"
        android:text="停止"
        android:id="@+id/button3"
        />
    <Button
        style="@style/buttonstyle"
        android:text="重置"
        android:id="@+id/button4"
```

```
            />
        </LinearLayout>
        <TextView
            android:layout_width="136dp"
            android:layout_height="wrap_content"
            android:textAppearance="?android:attr/textAppearanceSmall"
            android:text="计时显示"
            android:id="@+id/textView"
            android:layout_weight="0.42" />
</LinearLayout>
```

Styles.xml 文件的内容如下。

```
<?xml version="1.0" encoding="utf-8"?>
<?xml version="1.0" encoding="utf-8"?>
<resources>
    <style name="buttonstyle">
        <item name="android:layout_width">wrap_content</item>
        <item name="android:layout_height">wrap_content</item>
        <item name="android:layout_margin">10dp</item>
        <item name="android:textColor">#556600</item>
        <item name="android:textSize">24dp</item>
        <item name="android:layout_weight">0.02</item>
    </style>
</resources>
```

程序的控件布局结构如图 6.26 所示。

图 6.26 例 6-10 运行布局结构图

MainActivity.java 中的主要代码如下,主要是添加了样式。

```
import android.os.Bundle;
import android.os.SystemClock;
import android.support.v7.app.AppCompatActivity;
import android.view.View;
```

```java
import android.widget.Button;
import android.widget.Chronometer;
import android.widget.TextView;
public class MainActivity extends AppCompatActivity implements View.OnClickListener {
    Chronometer recordChronometer;
    Button startButton, stopButton, pauseButton, resetButton;
    TextView timer;
    boolean isPause=false;              //用于判断是否为暂停状态
    long recordingTime=0;
    @Override
    protected void onCreate(Bundle savedInstanceState){
        super.onCreate(savedInstanceState);
        setContentView(R.layout.activity_main);
        recordChronometer=(Chronometer)findViewById(R.id.chronometer);
        startButton=(Button)findViewById(R.id.button1);
        pauseButton=(Button)findViewById(R.id.button2);
        stopButton=(Button)findViewById(R.id.button3);
        resetButton=(Button)findViewById(R.id.button4);
        timer=(TextView)findViewById(R.id.textView);
        //recordChronometer.setFormat("计时:%s");
        startButton.setOnClickListener(this);
        pauseButton.setOnClickListener(this);
        stopButton.setOnClickListener(this);
        resetButton.setOnClickListener(this);
    }
    @Override
    public void onClick(View v){
        switch(v.getId()){
            case R.id.button1:
                //保证什么时候开始,计时都是从 0 开始
                recordChronometer.start();
                recordChronometer.setBase(SystemClock.elapsedRealtime());
                break;
            case R.id.button2:
                //实现了计时的暂停功能
                //如果不是处于暂停状态,则停止计时
                //如果处于暂停状态,则通过计算,从暂停的那个时间开始计时
                if(!isPause){
                    recordChronometer.stop();
                    isPause=true;
                    pauseButton.setText("继续计时");
                } else {
                    Double temp=Double.parseDouble(recordChronometer.getText()
                            .toString().split(":")[1]) * 1000;
                    recordChronometer
```

```
                .setBase((long)(SystemClock.elapsedRealtime()-temp));
            recordChronometer.start();
            pauseText();
        }
        break;
    case R.id.button4:
        //这句是为了防止单击暂停键后,chronometer处于stop()状态,
        //不能计时
        recordChronometer.start();
        recordChronometer.setBase(SystemClock.elapsedRealtime());

        break;
    case R.id.button3:
        //停止后,使时间归零
        recordChronometer.stop();
        recordChronometer.setBase(SystemClock.elapsedRealtime());
        pauseText();
        timer.setText("获得的时间为:"+recordChronometer.getText().toString());
        break;
    default:
        break;
    }
}
//设置处于暂停状态时,pause按钮的文字显示
public void pauseText(){
    if(isPause){
        pauseButton.setText("暂停计时");
        isPause=false;
    }
}
}
```

程序的运行结果如图 6.27 所示。

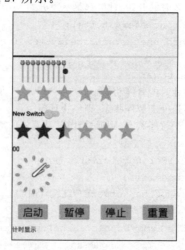

图 6.27　例 6-10 运行结果

6.8 项目实战：使用RadioButton实现主页底端导航条

6.8.1 项目分析

Android App肯定要由多个控件组成，也就是说，控件是一个程序的必要组成部分。CoffeeStore项目也不例外，本项目的首页是由一个ViewPager控件和多个单选按钮组成的，其中ViewPager控件在第7节介绍，本部分重点介绍单选按钮在首页中的应用。

在本项目中，计划利用单选按钮实现多个页面间的切换。由分析阶段确定的功能可知，共需要5个单选按钮，依次排列在页面底端，分别是"首页""购物车""分类""搜索""我的"，要实现的主页底端导航条效果如图6.28所示。

图6.28 首页的效果

6.8.2 项目实现

界面的下方包含5个单选按钮，接下来通过代码实现和应用。这部分的布局文件是activity_main.xml，代码如下。

```
<LinearLayout xmlns:android="http://schemas.android.com/apk/res/android"
    android:layout_width="match_parent"
    android:layout_height="match_parent"
    android:orientation="vertical">
    //定义viewpager控件
    <android.support.v4.view.ViewPager
        android:id="@+id/main_ViewPager"
        android:layout_width="match_parent"
        android:layout_height="0dp"
        android:layout_weight="1" />

    <RadioGroup
        android:id="@+id/main_tab_RadioGroup"
        android:layout_width="match_parent"
        android:layout_height="wrap_content"
        android:layout_gravity="bottom"
        android:gravity="center_vertical"
        android:orientation="horizontal">

        <RadioButton
            android:id="@+id/radio_home"
            android:layout_width="wrap_content"
```

```
            android:layout_height="wrap_content"
            android:checked="true"
            android:text="首页" />

        <RadioButton
            android:id="@+id/radio_shopcar"
            android:layout_width="wrap_content"
            android:layout_height="wrap_content"
            android:text="购物车" />

        <RadioButton
            android:id="@+id/radio_sort"
            android:layout_width="wrap_content"
            android:layout_height="wrap_content"
            android:text="分类" />

        <RadioButton
            android:id="@+id/radio_search"
            android:layout_width="wrap_content"
            android:layout_height="wrap_content"
            android:text="搜索" />

        <RadioButton
            android:id="@+id/radio_me"
            android:layout_width="wrap_content"
            android:layout_height="wrap_content"
            android:text="我的" />
    </RadioGroup>

</LinearLayout>
```

布局结构如图 6.29 所示。

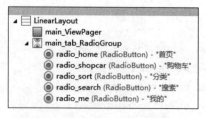

图 6.29　首页的布局结构图 1

通过单选按钮控制页面切换的代码将在第 7 章讲解。

上一章重点介绍了首页中应用的布局管理器,每个布局管理器的应用都是为了控制控件摆放的。所以首页中也涉及多个控件,本部分抽取一部分讲解。由图 6.30 可知,表格布局中排列了 ImageView 和 TextView，分别用来显示图片和文本。

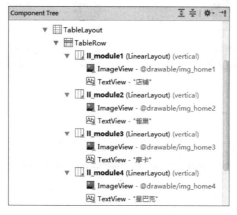

图 6.30　首页的布局结构图 2

ImageView 和 TextView 的属性设置如下。

```
<ImageView
    android:layout_width="50dp"
    android:layout_height="50dp"
    android:layout_gravity="center"
    android:padding="5dp"
    android:src="@drawable/img_home1" />
<TextView
    android:layout_width="match_parent"
    android:layout_height="wrap_content"
    android:gravity="center_horizontal"
    android:text="店铺"
    android:textColor="#929292"
    android:textSize="13sp" />
</LinearLayout>
```

6.8.3　项目说明

本项目用到了按钮类和文本类控件。应用单选按钮时，注意一定要把互斥的单选按钮放在一个组中。另外，本项目还应用了线性布局管理器和文本类控件。为了达到让文本类组件分两行排列的效果，还应用了 TableRow 组件，通过组件和布局管理器的结合应用，实现了本项目的首页界面，运行 Activity 文件显示的界面效果如图 6.31 和图 6.32 所示。

图 6.31 首页分类列表运行效果

图 6.32 首页单选按钮运行效果

6.9 知识扩展:创建和使用自定义控件

如果现有控件无法满足要实现的基本功能,就需要对现有控件的外观和(或)动作进行修改或扩展。Android 工具箱包含了很多创建 UI 必需的控件,但它们通常都是通用的,不能满足个性化的需求。通过扩展这些基本控件,可以对用户界面和功能做出改善。

要在一个已有控件的基础上创建一个新的 Widget,需要创建一个扩展了原有控件的新类。也可以通过组合多个控件来创建不可分割的、可重用的 Widget,从而使其可以综合使用多个相互关联的子控件的功能。创建复合控件时,必须对其布局、外观和它所包含的 View 之间的交互进行定义。复合控件是通过扩展一个 ViewGroup(通常是一个布局管理器)来创建的。因此,要创建一个新的复合控件,首先需要选择一个最适合放置子控件的布局类,再对其进行扩展。

【例 6-11】 自定义一个复合控件,由一个文本编辑框和用来清除其内容的按钮组成。

(1) 简单复合控件 Widget 的 XML 布局定义。

代码如下。

```
<?xml version="1.0" encoding="utf-8"?>
<LinearLayout xmlns:android="http://schemas.android.com/apk/res/android"
        android:layout_width="match_parent"
        android:layout_height="match_parent"
        android:orientation="vertical">
    <EditText
        android:id="@+id/editText"
        android:layout_width="match_parent"
        android:layout_height="wrap_content"
        />
    <Button
        android:id="@+id/clearbutton"
```

```
            android:layout_width="match_parent"
            android:layout_height="wrap_content"
            android:text="Clear"/>
</LinearLayout>
```

(2) 简单复合控件的定义。

代码如下。

```
import android.content.Context;
import android.util.AttributeSet;
import android.view.LayoutInflater;
import android.view.View;
import android.widget.Button;
import android.widget.EditText;
import android.widget.LinearLayout;
public class MainActivity extends LinearLayout {
    EditText editText;
    Button clearButton;
    public MainActivity(Context context, AttributeSet attrs){
        super(context, attrs);
        LayoutInflater inflater=LayoutInflater.from(context);
        View view=inflater.inflate(R.layout.layoutself, null);
        //另一种方式从布局资源中扩展View
        /* String infService=context.LAYOUT_INFLATER_SERVICE;
        LayoutInflater liu;
        liu= (LayoutInflater)getContext().getSystemService(infService);
        liu.inflate(R.layout.layoutself, this, true); */
        //获取对子控件的引用
        editText= (EditText)view.findViewById(R.id.editText);
        clearButton= (Button)view. findViewById(R.id.clearbutton);
        //链接这个功能
        addView(view);
        hookupButton();
    }
    private void hookupButton(){
        clearButton.setOnClickListener(new OnClickListener(){
            @Override
            public void onClick(View v){
                editText.setText(" ");
            }
        });
    }
    public MainActivity(Context context){
        super(context);
    }
```

}

(3) 自定义控件的使用。

创建了自己定制的控件后，就可以像使用其他任意的 Android 控件那样，在代码和布局中使用它们。要在一个布局中使用相同的控件，需要在布局定义中创建一个新的节点是指定完全限定的类名，如下面的 XML 代码段所示。

```xml
<?xml version="1.0" encoding="utf-8"?>
<LinearLayout xmlns:android="http://schemas.android.com/apk/res/android"
        android:layout_width="match_parent"
        android:layout_height="match_parent">
    <com.example.hp.ch06_11.MainActivity
        android:layout_width="match_parent"
        android:layout_height="wrap_content">
    </ com.example.hp.ch06_11.MainActivity >
</LinearLayout>
```

本章小结

Android 提供了大量控件，给平台开发者选择的空间，这也要求开发人员了解每一种控件及其使用方式，了解控件的各属性和方法的应用，根据实际情况选择最合适的控件。知识扩展环节介绍了如何创建和使用自定义控件，以满足现有控件不满足用户需求的情况。文中除了介绍 Android 中的常用控件，还结合前一章节所学的布局管理器给出了若干案例，用于巩固所学知识。

本章习题

1. AutoCompleteTextView 和 MultiAutoCompleteTextView 的区别是什么？
2. 如何给按钮控件实现事件处理？
3. 简述 ScrollView 的用法。
4. Android 中有哪些日期和时间类控件，它们各自的特点是什么？
5. 应用本章所学的文本框实现用户注册界面。
6. 应用本章所学的文本控件和星级控件实现项目中的咖啡评分功能，界面如图 6.33 所示。
7. 编写 Android 程序，使用文本类控件实现图 6.34 所示界面。
8. 编写 Android 程序，使用按钮控件实现图 6.35 所示界面，单击上方按钮时按钮变为红色，单击下方按钮时变为蓝色。
9. 编写 Android 程序，使用文本和按钮类控件实现图 6.36 所示界面，单击"提交"按钮，出现提示信息，显示文本中输入的姓名，以及选中的单选按钮和复选按钮信息。

第 6 章　Android 基本控件

图 6.33　咖啡评分界面

图 6.34　实现界面　　　　　　　　　　图 6.35　实现界面

图 6.36　实现界面

ViewPager 与 Fragment

本章概述

通过本章的学习,掌握 ViewPager 与 PagerAdapter、Fragment 的创建和应用、Intent 的创建和应用、Activity 与 Fragment 之间的交互,最后掌握 CoffeeStore 主页的实现方法。

学习重点与难点

重点:

(1) ViewPager 与 PagerAdapter。

(2) Fragment 的创建和应用。

(3) Intent 的创建和应用。

难点:

(1) Activity 与 Fragment 的综合应用。

(2) CoffeeStore 主页的实现。

学习建议

本章主要讲解 Android 中非常重要的 ViewPager 与 PagerAdapter 以及 Fragment、Intent 的应用。本章知识的广度增加了,读者应勤于练习,善于思考,多体会,多实践;课后延伸阅读,加深理解。

7.1　ViewPager 与 PagerAdapter

ViewPager 是 android-support-v4.jar 包中的一个系统控件,继承自 ViewGroup,专门用以实现左右滑动切换 View 的效果。ViewPager 的使用类似 ListView,需要有对应的 Adapter 进行数据绑定。

ViewPager 是一个抽象类,直接继承于 Object,要使用 ViewPager,首先要继承 PagerAdapter 类,至少要覆盖以下方法。

```
instantiateItem(ViewGroup, int)
destroyItem(ViewGroup, int, Object)
getCount()
isViewFromObject(View, Object)
```

每次创建 ViewPager 或滑动过程中,以上 4 个方法都会被调用,而 instantiateItem 和 destroyItem 中的方法要自己实现。

```
public abstract int getCount();
```

这个方法是获取当前窗体界面数。

```
public abstract boolean isViewFromObject(android.view.View arg0, java.lang.Object arg1);
```

这个方法用于判断是否由对象生成界面。

```
public java.lang.Object instantiateItem(android.view.View container, int position);
```

这个方法返回一个对象,这个对象表明了 PagerAdapter 适配器选择哪个对象放在当前的 ViewPager 中。

```
public void destroyItem(android.view.ViewGroup container, int position, java.lang.Object object);
```

这个方法是从 ViewGroup 中移出当前 View,类似 BaseAdapter,其中 instantiateItem 方法用来得到每个 View,destroyItem 用以控制某个 View 不需要时的回收处理。isViewFromObject 用来实现判断 View 和 Object 是否为同一个 View。

```
public class ViewPagerAdapter extends PagerAdapter{
    @Override
    public int getCount(){
        //TODO Auto-generated method stub
        return 0;
    }
    @Override
    public boolean isViewFromObject(View arg0, Object arg1){
        //TODO Auto-generated method stub
        return false;
    }
    @Override
    public void destroyItem(View container, int position, Object object){
        //TODO Auto-generated method stub
        super.destroyItem(container, position, object);
    }
    @Override
    public Object instantiateItem(View container, int position){
        //TODO Auto-generated method stub
        return super.instantiateItem(container, position);
    }
}
```

Fragment 和 ViewPager 经常结合应用,下面举例说明。

【例 7-1】 ViewPager 应用案例。

本案例程序能够显示图 7.1 所示的界面,接下来就进行详尽的分析。

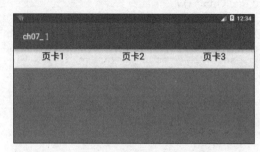

图 7.1 ViewPager 应用的效果图

activity_main.xml 是 Activity 的布局文件。

```xml
<?xml version="1.0" encoding="utf-8"?>
<LinearLayout xmlns:android="http://schemas.android.com/apk/res/android"
    android:layout_width="fill_parent"
    android:layout_height="fill_parent"
    android:orientation="vertical">
    <LinearLayout
        android:layout_width="match_parent"
        android:layout_height="45dp"
        android:orientation="horizontal">
        <TextView
            android:id="@+id/textView1"
            android:layout_width="fill_parent"
            android:layout_height="wrap_content"
            android:layout_weight="1.0"
            android:gravity="center"
            android:text="页卡 1"
            android:textAppearance="?android:attr/textAppearanceSmall"
            android:textColor="#000000"
            android:textSize="22.0dip" />
        <TextView
            android:id="@+id/textView2"
            android:layout_width="fill_parent"
            android:layout_height="wrap_content"
            android:layout_weight="1.0"
            android:gravity="center"
            android:text="页卡 2"
            android:textAppearance="?android:attr/textAppearanceSmall"
            android:textColor="#000000"
            android:textSize="22.0dip" />
```

```xml
    <TextView
        android:id="@+id/textView3"
        android:layout_width="fill_parent"
        android:layout_height="wrap_content"
        android:layout_weight="1.0"
        android:gravity="center"
        android:text="页卡 3"
        android:textAppearance="?android:attr/textAppearanceSmall"
        android:textColor="#000000"
        android:textSize="22.0dip" />
    </LinearLayout>
    <android.support.v4.view.ViewPager
        android:id="@+id/view"
        android:layout_width="wrap_content"
        android:layout_height="wrap_content" />
</LinearLayout>
```

用来切换的布局文件代码如下所示，共计 3 个布局文件。

```xml
<?xml version="1.0" encoding="utf-8"?>
<LinearLayout xmlns:android="http://schemas.android.com/apk/res/android"
    android:layout_width="fill_parent"
    android:layout_height="fill_parent"
    android:orientation="vertical"
    android:background="#158684" >
</LinearLayout>
<?xml version="1.0" encoding="utf-8"?>
<LinearLayout xmlns:android="http://schemas.android.com/apk/res/android"
    android:layout_width="fill_parent"
    android:layout_height="fill_parent"
    android:orientation="vertical"
    android:background="#150084" >
</LinearLayout>
<?xml version="1.0" encoding="utf-8"?>
<LinearLayout xmlns:android="http://schemas.android.com/apk/res/android"
    android:layout_width="fill_parent"
    android:layout_height="fill_parent"
    android:orientation="vertical"
    android:background="#158600" >
</LinearLayout>
```

MainActivity 是该示例主界面的 Activity，加载了 activity_main.xml 文件声明的界面布局。MainActivity.java 文件的完整代码如下。

```
import android.os.Bundle;
import android.support.v4.view.PagerAdapter;
```

```java
import android.support.v4.view.ViewPager;
import android.support.v7.app.AppCompatActivity;
import android.view.LayoutInflater;
import android.view.View;
import android.widget.ImageView;
import android.widget.TextView;
import java.util.ArrayList;
import java.util.List;
public class MainActivity extends AppCompatActivity {
    private ViewPager mPager;                    //页卡内容
    private List<View> listViews;                //Tab 页面列表
    private ImageView cursor;
    private TextView t1, t2, t3;                 //页卡头标
    /**
     * 初始化头标
     */
    private void InitTextView(){
        t1=(TextView)findViewById(R.id.textView1);
        t2=(TextView)findViewById(R.id.textView2);
        t3=(TextView)findViewById(R.id.textView3);
        t1.setOnClickListener(new MyOnClickListener(0));
        t2.setOnClickListener(new MyOnClickListener(1));
        t3.setOnClickListener(new MyOnClickListener(2));
    }
    public class MyOnClickListener implements View.OnClickListener {
        private int index=0;
        public MyOnClickListener(int i){
            index=i;
        }
        @Override
        public void onClick(View v){
            mPager.setCurrentItem(index);
        }
    }
    @Override
    protected void onCreate(Bundle savedInstanceState){
        super.onCreate(savedInstanceState);
        setContentView(R.layout.activity_main);
        mPager=(ViewPager)findViewById(R.id.view);
        listViews=new ArrayList<View>();
        LayoutInflater mInflater=getLayoutInflater();
        listViews.add(mInflater.inflate(R.layout.tab1, null));
        listViews.add(mInflater.inflate(R.layout.tab2, null));
        listViews.add(mInflater.inflate(R.layout.tab3, null));
```

```
        mPager.setAdapter(new MyPagerAdapter(listViews));
    }
    public class MyPagerAdapter extends PagerAdapter {
        public List<View>mListViews;
        public MyPagerAdapter(List<View>mListViews){
            this.mListViews=mListViews;
        }
        @Override
        public void destroyItem(View arg0, int arg1, Object arg2){
            ((ViewPager)arg0).removeView(mListViews.get(arg1));
        }
        @Override
        public int getCount(){
            return mListViews.size();
        }
        @Override
        public Object instantiateItem(View arg0, int arg1){
            ((ViewPager)arg0).addView(mListViews.get(arg1), 0);
            return mListViews.get(arg1);
        }
        @Override
        public boolean isViewFromObject(View arg0, Object arg1){
            return arg0 == (arg1);
        }
    }
}
```

7.2 Fragment 及其应用场合

Android 是在 3.0 版（API level 11）开始引入 Fragment 的，Fragment 的中文意思是碎片，引入 Fragment 是为了实现不同屏幕分辨率的动态和灵活 UI 设计。Fragment 与 Activity 很相似，用来在一个 Activity 中描述一些行为或一部分用户界面。可以把 Fragment 当做 Activity 中的模块，这个模块有自己的布局，有自己的生命周期，单独处理自己的输入，Activity 运行时可以加载或移除 Fragment 模块。Fragment 可以设计成在多个 Activity 中复用的模块。当开发的应用程序同时适用于平板电脑和手机时，可以利用 Fragment 实现灵活的布局，改善用户体验，如图 7.2 所示。

Fragment 的优点很多，包括以下几方面。
- 能够将 Activity 分离成多个可重用的组件，每个都有它自己的生命周期和 UI。
- 轻松地创建动态灵活的 UI 设计，对手机和平板电脑都有很好的适应性。
- Fragment 是一个独立的模块，紧紧地与 Activity 绑定在一起。可以在运行中动态地移除、加入、交换等。

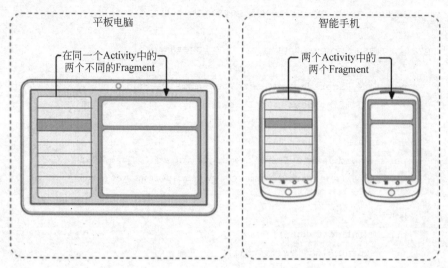

图 7.2 Fragment

- Fragment 提供一个新的方式,可以在不同的 Android 设备上统一 UI。
- Fragment 解决 Activity 间的切换不流畅问题,轻量切换。
- Fragment 替代 TabActivity 做导航,性能更好。
- Fragment 在 4.2 版本中新增嵌套 Fragment 使用方法,能够生成更好的界面效果。
- Fragment 做局部内容更新更方便,原来要把多个布局放到一个 Activity 里面,现在可以用多个 Fragment 来代替,只有在需要时才加载 Fragment,提高性能。

关于 Fragment,还有一点需要注意,就是每个 Fragment 都是有生命周期的。因为 Fragment 必须嵌入在 Acitivty 中使用,所以 Fragment 的生命周期和它所在的 Activity 是密切相关的。

如果 Activity 是暂停状态,其中所有的 Fragment 都是暂停状态;如果 Activity 是 stopped 状态,这个 Activity 中所有的 Fragment 都不能被启动;如果 Activity 被销毁,那么其中所有的 Fragment 都会被销毁。但是,当 Activity 在活动状态,可以独立控制 Fragment 的状态,比如加上或移除 Fragment。当这样进行 Fragment Transaction(转换)时,可以把 Fragment 放入 Activity 的堆栈中,这样就可以进行返回操作。Fragment 的生命周期如图 7.3 所示。

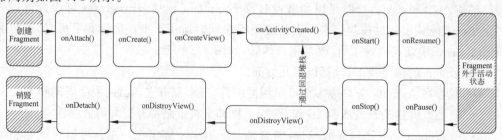

图 7.3 Fragment 的生命周期

管理 Fragment 的生命周期,大多数和管理 Activity 的生命周期很像,和 Activity 一样,Fragment 可以处于以下 3 种状态。

Resumed:运行中的 Activity 中的 Fragment 可见。

Paused:另一个 Activity 处于前台并拥有焦点,但是这个 Fragment 所在的 Activity 仍然可见(前台 Activity 局部透明或者没有覆盖整个屏幕)。

Stopped:或者是宿主 Activity 已经被停止,或者是 Fragment 从 Activity 被移除,但被添加到后台堆栈中。

停止状态的 Fragment 仍然活着(所有状态和成员信息被系统保持着)。然而,它对用户不再可见,如果 Activity 被系统回收,它也会被系统回收。

和 Activity 一样,可以使用 Bundle 保持 Fragment 的状态。如果 Activity 的进程被系统回收,并且当 Activity 被重新创建时,需要恢复 Fragment 状态时就可以用到,也可以在 Fragment 的 onSaveInstanceState() 期间保存状态,并可以在 onCreate()、onCreateView()或 onActivityCreated()期间恢复它。

在生命周期方面,Activity 和 Fragment 最重要的区别是各自如何在它的后台堆栈中储存。默认情况下,Activity 停止后,会被放到一个由系统管理的用于保存 Activity 的后台堆栈(因此用户可以使用 back 按键导航回退到它)。

然而,仅当在一个事务期间移除 Fragment 且显式调用 addToBackStack()请求保存实例时,才被放到一个由宿主 Activity 管理的后台堆栈。

7.3 创建 Fragment

使用 Fragment 时,需要继承 Fragment 或者 Fragment 的子类(DialogFragment,ListFragment,PreferenceFragment,WebViewFragment),所以 Fragment 的代码看起来和 Activity 的类似。除了继承以外,至少应实现 onCreate()、onCreateView()和 onPause() 3 个生命周期的回调函数。

- onCreate()函数是在 Fragment 创建时被调用,用来初始化 Fragment 中的必要组件。
- onCreateView()函数是 Fragment 在用户界面上第一次绘制时被调用,并返回 Fragment 的根布局视图。
- onPause()函数是在用户离开 Fragment 时被调用,用来保存 Fragment 中用户输入或修改的内容。

如果仅通过 Fragment 显示元素,而不进行任何数据保存和界面事件处理,则仅可实现 onCreateView()函数。

例如,需要创建一个名称为 NewsFragment 的子类,并重写 onCreateView()方法,可以使用如下代码。

```
public class NewsFragment extends Fragment {
    public NewsFragment(){
        //Required empty public constructor
```

```
    }
    @Override
    public View onCreateView(LayoutInflater inflater, ViewGroup container,
            Bundle savedInstanceState){
        return inflater.inflate(R.layout.news, container, false);
    }
}
```

系统首次调用 Fragment 时,如果想绘制一个 UI 界面,那么在 Fragment 中必须重写 onCreateView()方法,返回一个 View;如果 Fragment 没有 UI 界面,可以返回 null。

下面通过一个例子来了解 Fragment 的使用。

【例 7-2】 Fragment 应用案例。

本案例中共有 3 个 Java 文件,其中有 2 个 Fragment 类的子类,1 个是 Activity 的子类,程序能够显示图 7.4 中的(b)界面,接下来就进行详尽的分析。

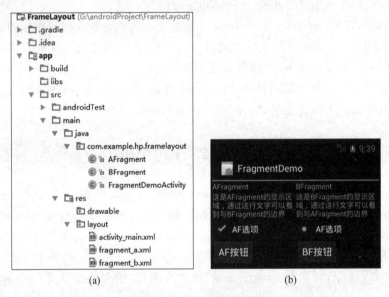

图 7.4 Fragment 的案例图

activity_main.xml 文件是 Activity 的布局文件,两个 Fragment 在界面上的位置关系就在这个文件中定义。activity_main.xml 的代码如下。

```
<LinearLayout xmlns:android="http://schemas.android.com/apk/res/android"
    android:orientation="horizontal"
    android:layout_width="match_parent"
    android:layout_height="match_parent">
<fragment android:name ="edu.hrbeu.FragmentDemo.AFragment"
        android:id="@+id/fragment_a"
        android:layout_weight="1"
        android:layout_width="0px"
        android:layout_height="match_parent" />
```

```xml
<fragment android:name = "edu.hrbeu.FragmentDemo.BFragment"
        android:id="@+id/fragment_b"
        android:layout_weight="1"
        android:layout_width="0px"
        android:layout_height="match_parent" />
</LinearLayout>
```

FragmentDemoActivity是该示例主界面的Activity,加载了activity_main.xml文件声明的界面布局。FragmentDemoActivity.java文件的完整代码如下。

```java
import android.os.Bundle;
import android.app.Activity;
import android.view.Menu;
public class FragmentDemoActivity extends Activity {
    @Override
    protected void onCreate(Bundle savedInstanceState){
        super.onCreate(savedInstanceState);
        setContentView(R.layout.activity_main);
    }
    @Override
    public boolean onCreateOptionsMenu(Menu menu){
        //Inflate the menu; this adds items to the action bar if it is present.
        getMenuInflater().inflate(R.menu.menu_main, menu);
        return true;
    }
}
```

Android系统会根据代码setContentView(R.layout.main);的内容加载界面布局文件activity_main.xml,然后通过activity_main.xml文件中对Fragment所在的"包+类"的描述,找到Fragment的实现类,并调用类中的onCreateView()函数绘制界面元素。

AFragment.java文件的核心代码如下。

```java
public class AFragment extends Fragment{
    @Override
    public View onCreateView(LayoutInflater inflater, ViewGroup container,
                    Bundle savedInstanceState){
        //TODO Auto-generated method stub
        return inflater.inflate(R.layout.fragment_a, container,false);
    }
}
```

AFragment中只实现了onCreateView()函数,返回值是AFragment的视图,代码return inflater.inflate(R.layout.frag_a, container, false);使用inflate()函数,通过指定资源文件R.layout.frag_a获取到AFragment的视图。

BFragment.java文件的核心代码如下。

```
public class BFragment extends Fragment{
    @Override
    public View onCreateView(LayoutInflater inflater, ViewGroup container,
                    Bundle savedInstanceState){
        //TODO Auto-generated method stub
        return inflater.inflate(R.layout.fragment_b, container, false);
    }
}
```

最后,给出 fragment_a.xml 文件的全部代码如下。

```xml
<?xml version="1.0" encoding="utf-8"?>
<LinearLayout xmlns:android="http://schemas.android.com/apk/res/android"
    android:layout_width="wrap_content"
    android:layout_height="wrap_content"
    android:orientation="vertical" >
<TextView
        android:layout_width="wrap_content"
        android:layout_height="wrap_content"
        android:text="AFragment" />
<TextView
        android:layout_width="wrap_content"
        android:layout_height="wrap_content"
android:text="这是 AFragment 的显示区域,通过这行文字可以看到与 BFragment 的边界" />
<CheckBox
        android:layout_width="wrap_content"
        android:layout_height="wrap_content"
        android:text="AF 选项" />
<Button
        android:layout_width="wrap_content"
        android:layout_height="wrap_content"
        android:text="AF 按钮" />
</LinearLayout>
```

fragment_b.xml 文件的全部代码如下。

```xml
<?xml version="1.0" encoding="utf-8"?>
<LinearLayout xmlns:android="http://schemas.android.com/apk/res/android"
    android:layout_width="wrap_content"
    android:layout_height="wrap_content"
    android:orientation="vertical" >
    <TextView
        android:layout_width="wrap_content"
        android:layout_height="wrap_content"
        android:text="BAFragment" />
    <TextView
```

```
        android:layout_width="wrap_content"
        android:layout_height="wrap_content"
        android:text="这是 BFragment 的显示区域,通过这行文字可以看到与 AFragment
        的边界" />
    <CheckBox
        android:layout_width="wrap_content"
        android:layout_height="wrap_content"
        android:text="BF 选项" />
    <Button
        android:layout_width="wrap_content"
        android:layout_height="wrap_content"
        android:text="BF 按钮" />
</LinearLayout>
```

Fragment 也可以通过代码动态添加,下面通过一个例子来了解 Fragment 的动态添加。

【例 7-3】 Fragment 的动态添加应用案例。

本案例程序能够显示图 7.5 所示的界面,下面对案例进行详尽的分析。

图 7.5　Fragment 动态添加的案例图

activity_main.xml 文件是 Activity 的布局文件,Fragment 将显示在 container 组件中。

```
<RelativeLayout xmlns:android="http://schemas.android.com/apk/res/android"
    xmlns:tools="http://schemas.android.com/tools"
    android:layout_width="match_parent"
    android:layout_height="wrap_content"
    android:paddingBottom="@dimen/activity_vertical_margin"
    android:paddingLeft="@dimen/activity_horizontal_margin"
    android:paddingRight="@dimen/activity_horizontal_margin"
    android:paddingTop="@dimen/activity_vertical_margin"
    tools:context=".MainActivity" >
    <LinearLayout
        android:id="@+id/container"
        android:layout_width="match_parent"
        android:layout_height="wrap_content"
        android:orientation="vertical" >
    </LinearLayout>
</RelativeLayout>
```

Fragment 布局文件的代码如下。

```xml
<FrameLayout xmlns:android="http://schemas.android.com/apk/res/android"
    xmlns:tools="http://schemas.android.com/tools"
    android:layout_width="match_parent"
    android:layout_height="match_parent"
    tools:context=".ContentFragment" >
    <TextView
        android:layout_width="match_parent"
        android:layout_height="match_parent"
        android:text="这是 ContentFragment 界面" />
</FrameLayout>
```

MainActivity 是该示例主界面的 Activity，加载了 activity_main.xml 文件声明的界面布局。MainActivity.java 文件的完整代码如下。

```java
import android.app.FragmentManager;
import android.app.FragmentTransaction;
import android.os.Bundle;
import android.support.v7.app.AppCompatActivity;
public class MainActivity extends AppCompatActivity {
    @Override
    protected void onCreate(Bundle savedInstanceState){
        super.onCreate(savedInstanceState);
        setContentView(R.layout.activity_main);
        FragmentManager fragmentManager=this.getFragmentManager();
        FragmentTransaction tx=fragmentManager.beginTransaction();
        //第一个参数是 activity_main 布局文件中的 linearlayout 的 id,第二个参数是
            Framgment 的类名
        tx.add(R.id.container1, new ContentFragment());
        tx.commit();
    }
}
```

Fragment 的界面代码如下。

```java
import android.app.Fragment;
import android.os.Bundle;
import android.util.Log;
import android.view.LayoutInflater;
import android.view.View;
import android.view.ViewGroup;
public class ContentFragment extends Fragment {
    public ContentFragment(){
        //Required empty public constructor
    }
    @Override
    public View onCreateView(LayoutInflater inflater, ViewGroup container,
```

```
                Bundle savedInstanceState){
    //Inflate the layout for this fragment
    Log.v("ContentFragment", "onCreateView");
    return inflater.inflate(R.layout.fragment_content, container, false);
    }
}
```

7.4 初识 Intent

　　一个 Android 应用主要是由 Activity、Intent、Service、Content Provider 这 4 种组件组成的。而这 4 种组件是独立的，它们之间可以互相调用，协调工作，最终组成一个真正的 Android 应用。这些组件之间的通讯主要是由 Intent 协助完成的。Intent 负责对应用中一次操作的动作、动作涉及数据、附加数据进行描述，Android 则根据此 Intent 的描述，负责找到对应的组件，将 Intent 传递给调用的组件，并完成组件的调用。因此，Intent 在这里起着一个媒体中介的作用，专门提供组件互相调用的相关信息。

　　例如，在一个联系人维护的应用中，当在一个联系人列表屏幕（假设对应的 Activity 为 listActivity）上单击某个联系人后，希望能够跳出此联系人的详细信息屏幕（假设对应的 Activity 为 detailActivity）。为了实现这个目的，listActivity 需要构造一个 Intent，用于告诉系统要做"查看"动作，此动作对应的查看对象是"某联系人"，然后调用 startActivity (Intent intent)，传入构造的 Intent，系统会根据此 Intent 中的描述到 Manifest 中找到满足此 Intent 要求的 Activity，系统会调用找到的 Activity，即 detailActivity，最终传入 Intent，detailActivity 则会根据此 Intent 中的描述执行相应的操作。

7.4.1 Intent 对象的基本概念

　　一个 Android 程序由多个组件组成，各个组件之间使用 Intent 通信，Intent 可翻译为意图，Intent 对象中包含组件名称、动作、数据等内容。根据 Intent 中的内容，Android 系统可以启动需要的组件。

　　Intent 是一种在不同组件之间传递的请求信息，是应用程序发出的请求和意图。作为一个完整的消息传递机制，Intent 不仅需要发送端，还需要接收端。

　　理解 Intent 的关键之一是理解清楚 Intent 的两种基本用法：一种是显式的 Intent，即在构造 Intent 对象时就指定接收者；另一种是隐式的 Intent，即 Intent 的发送者在构造 Intent 对象时并不知道也不关心接收者是谁，有利于降低发送者和接收者之间的耦合。

　　对于显式 Intent，Android 不需要进行解析，因为目标组件已经很明确，Android 需要解析的是那些隐式 Intent，通过解析，将 Intent 映射给可以处理此 Intent 的 Activity、IntentReceiver 或 Service。

　　Intent 解析机制主要是通过查找已注册在 AndroidManifest.xml 中的所有 IntentFilter 及其中定义的 Intent，最终找到匹配的 Intent。在这个解析过程中，Android 是通过 Intent 的 action、type、category 这 3 个属性来进行判断的，判断方法如下：

如果 Intent 指明定了 action，则目标组件的 IntentFilter 的 action 列表中就必须包含这个 action，否则不能匹配。

如果 Intent 没有提供 type，系统将从 data 中得到数据类型。和 action 一样，目标组件的数据类型列表中必须包含 Intent 的数据类型，否则不能匹配。

如果 Intent 中的数据不是 content：类型的 URI，而且 Intent 也没有明确指定它的 type，将根据 Intent 中数据的 scheme（比如 http：或者 mailto：）进行匹配。同上，Intent 的 scheme 必须出现在目标组件的 scheme 列表中。

如果 Intent 指定了一个或多个 category，这些类别必须全部出现在组建的类别列表中。比如 Intent 中包含了两个类别：LAUNCHER_CATEGORY 和 ALTERNATIVE_CATEGORY，解析得到的目标组件必须至少包含这两个类别。

7.4.2　Intent 对象的基本使用方法

Intent 有显式和隐式之分，显式的 Intent 是根据组件的名称直接启动要启动的组件，例如 Service 或者 Activity；隐式的 Intent 通过配置 Action、Category、Data 来找到匹配的组件并启动。

显式启动例子如下。

```
Intent intent=new Intent(IntentDemo.this, ActivityToStart.class);
startActivity(intent);
```

隐式启动例子如下。

```
Intent intent=new Intent(Intent.ACTION_VIEW, Uri.parse("http://www.google.com.hk"));
startActivity(intent);
```

指定了 component 属性后，就是显式地指定了目标组件，也就是接收端。如果没有明确指定目标组件，那么 Android 系统会使用 Intent 里的 action、data、category 3 个属性来寻找和匹配接收端。

使用 Intent 对象中，需要指定 Intent 目标组件的组件名称。通常，组件是一个 ComponentName 对象，由目标组件的完全限定类名和组件所在应用程序配置文件中设置的报名组合而成。组件名称的报名部分和配置文件中设置的报名不必匹配。

组件名称是可选的，如果设置，Intent 对象会被发送给指定类的实例；如果没有设置，Android 使用 Intent 对象中的其他信息决定合适的目标。

组件名称可以使用 setComponent()、setClass() 或 SetClassName() 方法设置，使用 getComponent() 方法读取。

Intent 是对执行某个操作的一个抽象描述，描述的内容包括对执行动作 Action 的描述。Action 是个字符串，是对将执行的动作的描述，动作包括标准 Activity 动作和标准广播动作，Intent 类中定义了一些字符串常量作为标准动作，如表 7.1 所示。

表 7.1　Activity 动作和标准广播动作表

常　　量	类型	说　　明
ACTION_CALL	activity	初始化一个电话呼叫
ACTION_EDIT	activity	显示用户要编辑的数据
ACTION_MAIN	activity	将该 Activity 作为 task 的第一个 Activity,没有数据输入和返回
ACTION_SYNC	activity	在设备上同步服务器上的数据
ACTION_BATTERY_LOW	broadcast receiver	电量不足的警告
ACTION_HEADSET_PLUG	broadcast receiver	耳机插入设备,或者从设备中拔出
ACTION_SCREEN_ON	broadcast receiver	屏幕已经点亮
ACTION_TIMEZONE_CHANGED	broadcast receiver	时区设置改变

除了预定义的动作外,开发人员也可以自定义动作字符串来启动应用程序中的组件。使用自定义的动作时,需要加上包名作为前缀,如 com.example.project.SHOW_COLOR,并可定义相应的 Activity 来处理用户的自定义动作。

动作很大程度上决定了 Intent 其他部分的组成,特别是数据(data)和额外(extras)部分,就像方法名称决定了参数和返回值。因此,动作名称越具体越好,并且将它与 Intent 其他部分紧密联系。也就是开发人员应该为组建能处理的 Intent 对象定义完整的协议,而不是单独定义一个动作。

Intent 对象中的动作使用 setAction()方法设置,使用 getAction()方法读取。

使用 Intent 对象时,另外一个重要的组成部分是 Data,也就是执行动作要操作的数据。Data 表示操作数据的 URI 和 MIME 类型,不同动作与不同类型的数据规范匹配。例如,如果动作是 ACTION_EDIT,数据应该是包含用来编辑的文档的 URI;如果动作是 ACTION_CALL,数据应该是包含呼叫号码的 tel:URI。类似地,如果动作是 Action_View 而且数据是 http:URI,接收的 Activity 用来下载和显示 URI 指向的数据。

在将 Intent 与处理它的数据的组件匹配时,除了数据的 URI,也有必要了解其MIME 类型。例如,能够显示图片数据的组件不应用来播放音频文件。

在多数情况下,数据类型可以从 URI 中推断,尤其是 contend:URI。它表示数据存在于设备上并由 ContentProvider 控制,但是,类型信息也可以显式地设置在 Intent 对象中。setData()方法仅能指定数据的 URI,setType()方法仅能指定数据的 MIME 类型,setDataAndType()方法可以同时设置 URI 和 MIME 类型,使用 getData()方法可以读取URI,使用 getType()方法可以读取类型。

type(数据类型),显式指定 Intent 的数据类型(MIME)。一般 Intent 的数据类型能够根据数据本身进行判定,但是通过设置这个属性,可以强制采用显式指定的类型,而不再进行推导。

category(类别)，被执行动作的附加信息。例如 LAUNCHER_CATEGORY 表示 Intent 的接受者应该在 Launcher 中作为顶级应用出现；而 ALTERNATIVE_CATEGORY 表示当前的 Intent 是一系列可选动作中的一个，这些动作可以在同一块数据上执行。还有其他附加信息，如表 7.2 所示。

表 7.2 Activity 动作表

动 作	含 义
CATEGORY_BROWSABLE	目标 Activity 可以使用浏览器显示数据
CATEGORY_GADGET	该 Activity 可以被包含在另外一个装载小工具的 Activity 中
CATEGORY_HOME	设置当前 Activity 是当用户按下 Home 键时见到的屏幕
CATEGORY_LAUNCHER	设置第一个启动的 Activity
CATEGORY_PREFERENCE	该 Activity 是一个选项面板

extras(附加信息)，是其他所有附加信息的集合。使用 extras 可以为组件提供扩展信息，比如，如果要执行"发送电子邮件"这个动作，可以将电子邮件的标题、正文等保存在 extras 里，传给电子邮件发送组件。

7.4.3 使用 Intent 对象在 Activity 之间传递数据

Intent 的主要作用就是在 Android 组件中传递数据，下面讲解使用 Intent 对象在 Activity 之间传递数据的第 1 种方法。

在起始 Activity 中发送数据，代码如下。

```
protected void onCreate(Bundle saveInstanceState){
    super.onCreate(saveInstanceState);
    setContentView(R.layout.thisactivity);
    Intent intent=new Intent();
    //设置起始 Activity 和目标 Activity,表示数据从这个 Activity 传到下个 Activity
    intent.setClass(ThisActivity.this,TargetActivity.class);
    //绑定数据
    intent.putExtra("username",username);         //也可以绑定数组
    intent.putExtra("userpass",userpass);
    //打开目标 Activity
    startActivity(intent);
}
```

在目标 Activity 中接收数据，代码如下。

```
protected void onCreate(Bundle saveInstanceState){
    super.onCreate(saveInstanceState);
    setContentView(R.layout.targetactivity);
    //获得意图
    Intent intent=getIntent();
```

```
    //读取数据
    String name=intent.getStringExtra("username");
    String pass=intent.getStringExtra("userpass");
}
```

使用 Intent 对象在 Activity 之间传递数据的第 2 种方法借助 Bundle 实现。Bundle 其实就是一个 Key-Value 的映射，Intent 描述数据的 putExtra 方法采用的 Key-Value 形式。其实 Intent 在内部定义的事后就有个 Bundle 类型的成员变量，putExtra 方法就是将传进来的参数放到这个 Bundle 变量。所以 Bundle 有许多类似 putExtra 的方法来存放数据。

在起始 Activity 中发送数据，代码如下。

```
protected void onCreate(Bundle saveInstanceState){
    super.onCreate(saveInstanceState);
    setContentView(R.layout.thisactivity);
    Intent intent=new Intent();
    //设置起始 Activity 和目标 Activity,表示数据从这个 Activity 传到下个 Activity
    intent.setClass(ThisActivity.this,TargetActivity.class);
    //一次绑定多个数据
    Bundle bundle=new Bundle();
    bundle.putString("username",username);
    bundle.putString("userpass",userpass);
    intent.putExtras(bundle);
    //打开目标 Activity
    startActivity(intent);
}
```

在目标 Activity 中接收数据，代码如下。

```
protected void onCreate(Bundle saveInstanceState){
    super.onCreate(saveInstanceState);
    setContentView(R.layout.targetactivity);
    //获得意图
    Intent intent=getIntent();
    //读取数据
    Bundle bundle=intent.getExtras();
    String name=bundle.getString("username");
    String pass=bundle.getString("userpass");
}
```

当需要从目标 Activity 回传数据到原 Activity 时，可以使用上述方法定义一个新的 Intent 来传递数据，也可以使用 startActivityForResult(Intent intent，int requestCode); 方法。

在起始 Activity 中发送数据，代码如下。

```java
protected void onCreate(Bundle saveInstanceState){
    super.onCreate(saveInstanceState);
    setContentView(R.layout.thisactivity);
    Intent intent=new Intent();
    //设置起始 Activity 和目标 Activity,表示数据从这个 Activity 传到下个 Activity
    intent.setClass(ThisActivity.this,TargetActivity.class);
    //绑定数据
    intent.putExtra("username",username);              //也可以绑定数组
    intent.putExtra("userpass",userpass);
    //打开目标 Activity
    startActivityForResult(intent,1);
}
//需要重写 onActivityResult 方法
protected void onActivityResult(int requestCode, int resultCode, Intent intent){
    super.onActivityResult(requestCode,resultCode,intent);
    //判断结果码是否与回传的结果码相同
    if(resultCode ==1){
        //获取回传数据
        String name=intent.getStringExtra("name");
        String pass=intent.getStringExtra("pass");
        //对数据进行操作
        ...
    }
}
```

在目标 Activity 中接收数据,代码如下。

```java
protected void onCreate(Bundle saveInstanceState){
    super.onCreate(saveInstanceState);
    setContentView(R.layout.targetactivity);
    //获得意图
    Intent intent=getIntent();
    //读取数据
    String name=intent.getStringExtra("username");
    String pass=intent.getStringExtra("userpass");
    //从 EditText 中获取新的数据给 name 和 pass
    name=editText1.getText().toString();
    pass=editText2.getText().toString()
    //数据发生改变,需要把改变后的值传递回原来的 Activity
    intent.putExtra("name",name);
    intent.putExtra("pass",pass);
    //setResult(int resultCode,Intent intent)方法
    setResult(1,intent);
    //销毁此 Activity,摧毁此 Activity 后将自动回到上一个 Activity
    finish();
}
```

7.5 Activity 与 Fragment 之间的交互

Fragment 组件使用时必然会与 Activity 进行交互，下面介绍 Activity 与 Fragment 之间的交互和生命周期的关系。前面已介绍过 Fragment 是依赖于 Activity 的，而且生命周期也跟 Activity 绑定一起。

7.5.1 为 Activity 创建事件回调方法

在一些情况下，可能需要一个 Fragment 与 Activity 分享事件。一个好的方法是在 Fragment 中定义一个回调的 Interface，并要求宿主 Activity 实现它。当 Activity 通过 Interface 接收到一个回调，必要时可以和在 Layout 中的其他 Fragment 分享信息。例如，如果一个新的应用在 Activity 中有 2 个 Fragment，一个显示文章列表（Fragment A），另一个显示文章内容（Fragment B），那么 Fragment A 必须负责通知 Activity 哪一个文章列表被选中，然后根据选中的内容进行 Fragment B 的显示。

在这个例子中，OnArticleSelectedListener 接口在 Fragment A 中声明以下代码。

```
public static class FragmentA extends ListFragment {
    //...
    //Container Activity must implement this interface
    public interface OnArticleSelectedListener {
        public void onArticleSelected(Uri articleUri);
    }
}
```

然后 Fragment 的宿主 Activity 实现 OnArticleSelectedListener 接口，并覆写 onArticleSelected()来通知 Fragment B 处理 Fragment A 生成的事件。为了确保宿主 Activity 实现这个接口，Fragment A 的 onAttach()回调方法（添加 Fragment 到 Activity 时由系统调用）通过将作为参数传入 onAttach()的 Activity 做类型转换来实例化一个 OnArticleSelectedListener 实例。

在 Fragment A 中添加 onAttach()方法的代码如下。

```
public static class FragmentA extends ListFragment {
    OnArticleSelectedListener mListener;
//...
    @Override
    public void onAttach(Activity activity){
        super.onAttach(activity);
        try {
            mListener= (OnArticleSelectedListener)activity;
        } catch(ClassCastException e){
            throw new ClassCastException(activity.toString()+" must
            implementOnArticleSelectedListener");
```

```
        }
    }
    //...
}
```

如果 Activity 没有实现接口，Fragment 会抛出 ClassCastException 异常。正常情形下，mListener 成员会保持一个到 Activity 的 OnArticleSelectedListener 实现的引用，因此 Fragment A 可以通过调用在 OnArticleSelectedListener 接口中定义的方法分享事件给 Activity。例如，如果 Fragment A 是一个 ListFragment 的子类，每次用户单击一个列表项，系统会调用 Fragment 中的 onListItemClick()，然后由后者调用 onArticleSelected() 来分配事件给 Activity。

在 Fragment A 中添加 onListItemClick() 方法的代码如下。

```
public static class FragmentA extends ListFragment {
    OnArticleSelectedListener mListener;
    //...
    @Override
    public void onListItemClick(ListView l, View v, int position, long id){
        //Append the clicked item's row ID with the content provider Uri
        Uri noteUri = ContentUris.withAppendedId(ArticleColumns.CONTENT_URI, id);
        //Send the event and Uri to the host activity
        mListener.onArticleSelected(noteUri);
        //...
    }
}
```

传给 onListItemClick() 的 id 参数是被单击的项的行 ID，Activity（或其他 Fragment）用来从应用的 ContentProvider 获取文章。

7.5.2 添加项目到 ActionBar

用户的 Fragment 可以通过实现 onCreateOptionMenu() 提供菜单项给 Activity 的选项菜单（以此类推，Action Bar 也一样）。为了使这个方法接收调用，无论如何，必须在 onCreate() 期间调用 setHasOptionsMenu() 来指出 Fragment 愿意添加 Item 到选项菜单（否则，Fragment 将接收不到对 onCreateOptionsMenu() 的调用）。

随后从 Fragment 添加到 Option 菜单的任何项，都会被追加到现有菜单项的后面。当一个菜单项被选择，Fragment 也会接收到对 onOptionsItemSelected() 的回调。也可以在 Fragment Layout 中通过调用 registerForContextMenu() 注册一个 View 来提供一个环境菜单。打开环境菜单，Fragment 接收到一个对 onCreateContextMenu() 的调用；而当用户选择一个项目，Fragment 会接收到一个对 onContextItemSelected() 的调用。

需要注意的是，尽管 Fragment 会接收到它所添加的每一个菜单项被选择后的回调，但实际选择一个菜单项时，Activity 会首先接收到对应的回调。如果 Activity 的 on-item-selected 回调函数实现并没有处理被选中的项目，事件才会被传递到 Fragment 的回调。这个规则适用于选项菜单和环境菜单。

7.5.3 与 Activity 生命周期的协调工作

Fragment 所生存的 Activity 的生命周期,直接影响 Fragment 的生命周期。每一个 Activity 生命周期的回调行为都会引起每一个 Fragment 中类似的回调。

例如,当 Activity 接收到 onPause()时,其中的每一个 Fragment 都会接收到 onPause()。Fragment 有几个额外的生命周期回调方法,用来处理与 Activity 的交互,以便执行诸如创建和销毁 Fragment 的 UI 的动作。这些额外的回调方法如下。

onAttach():当 Fragment 被绑定到 Activity 时被调用(Activity 会被传入)。
onCreateView():创建和 Fragment 关联的 View Hierarchy 时调用。
onActivityCreated():当 Activity 的 onCreate()方法返回时被调用。
onDestroyView():当和 Fragment 关联的 View Hierarchy 正在被移除时调用。
onDetach():当 Fragment 从 Activity 解除关联时被调用。

7.6 项目实战:CoffeeStore 主页滑动功能的实现

7.6.1 项目分析

前面已经介绍了 CoffeeStore 项目,本节重点介绍如何使用本篇章所学知识设计 CoffeeStore 主页,实现主页内容的滑动和切换功能。

首先介绍首页面的实现。首页是由一个 ViewPager 控件和多个单选按钮组成的,其中 ViewPager 控件将容纳多个 Fragment,具体展示哪个 Fragment,是由单选按钮控制的。

7.6.2 项目实现

首先创建 5 个 Fragment,如图 7.6 所示。

然后为每个 Fragment 创建一个对应的布局文件。接着创建主页的 MainActivity,在其对应的布局文件中使用 ViewPager 控件容纳 5 个 Fragment,在下方放置 5 个 RadioButton,用户既可以通过滑动来切换 5 个页面,也可以通过单击单选按钮来切换 5 个页面。主页对应的布局代码如下。

图 7.6 主页需要创建的 5 个 Fragment

```
<LinearLayout xmlns:android="http://schemas.android.com/apk/res/android"
    android:layout_width="fill_parent"
    android:layout_height="fill_parent"
    android:orientation="vertical" >
    <android.support.v4.view.ViewPager
        android:layout_width="fill_parent"
        android:layout_height="0dp"
```

```xml
        android:layout_weight="1"
        android:id="@+id/main_ViewPager"/>
    <RadioGroup
        android:id="@+id/main_tab_RadioGroup"
        android:layout_width="match_parent"
        android:layout_height="wrap_content"
        android:layout_gravity="bottom"
        android:background="@drawable/maintab_toolbar_bg"
        android:gravity="center_vertical"
        android:orientation="horizontal"
        >
        <RadioButton
            android:id="@+id/radio_home"
            style="@style/main_tab"
            android:drawableTop="@drawable/icon_index"
            android:text="首页"
            android:checked="true"
            />
        <RadioButton
            android:id="@+id/radio_shopcar"
            style="@style/main_tab"
            android:drawableTop="@drawable/icon_shopcar"
            android:text="购物车"
            />
        <RadioButton
            android:id="@+id/radio_sort"
            style="@style/main_tab"
            android:drawableTop="@drawable/icon_sort"
            android:text="分类"
            />
        <RadioButton
            android:id="@+id/radio_search"
            style="@style/main_tab"
            android:drawableTop="@drawable/icon_search"
            android:text="搜索"
            />
        <RadioButton
            android:id="@+id/radio_me"
            style="@style/main_tab"
            android:drawableTop="@drawable/icon_me"
            android:text="我的"
            />
    </RadioGroup>
</LinearLayout>
```

其中,fragment_home 的布局文件代码如下。

```xml
<?xml version="1.0" encoding="utf-8"?>
<ScrollView xmlns:android="http://schemas.android.com/apk/res/android"
    android:id="@+id/scrollView1"
    android:layout_width="fill_parent"
    android:layout_height="fill_parent" >
<LinearLayout
    android:layout_width="match_parent"
    android:layout_height="match_parent"
    android:orientation="vertical"
  >
    <FrameLayout
        android:id="@+id/framelayout"
        android:layout_width="match_parent"
        android:layout_height="150dp"
        android:background="#ffffff" >
        <org.taptwo.android.widget.ViewFlow
            android:id="@+id/viewflow"
            android:layout_width="match_parent"
            android:layout_height="match_parent" >
        </org.taptwo.android.widget.ViewFlow>
        <LinearLayout
            android:layout_width="match_parent"
            android:layout_height="wrap_content"
            android:layout_gravity="bottom"
            android:background="#88252525"
            android:gravity="center"
            android:padding="3dp" >
            <org.taptwo.android.widget.CircleFlowIndicator
                android:id="@+id/viewflowindic"
                android:layout_width="wrap_content"
                android:layout_height="wrap_content"
                android:layout_gravity="center_horizontal|bottom"
                android:padding="2dp"
                />
        </LinearLayout>
    </FrameLayout>
    <TableLayout
        android:layout_width="match_parent"
        android:layout_height="wrap_content"
        android:background="@color/page_background"
        android:orientation="vertical"
        android:stretchColumns="*" >
```

```xml
<TableRow
    android:layout_width="match_parent"
    android:layout_height="wrap_content"
    android:paddingTop="6dp" >
    <LinearLayout
        android:id="@+id/ll_module1"
        android:layout_width="wrap_content"
        android:layout_height="wrap_content"
        android:orientation="vertical" >
        <ImageView
            android:layout_width="50dp"
            android:layout_height="50dp"
            android:layout_gravity="center"
            android:padding="5dp"
            android:src="@drawable/img_home1" />
        <TextView
            android:layout_width="match_parent"
            android:layout_height="wrap_content"
            android:gravity="center_horizontal"
            android:text="店铺"
            android:textColor="#929292"
            android:textSize="13sp" />
    </LinearLayout>
    <LinearLayout
        android:id="@+id/ll_module2"
        android:layout_width="wrap_content"
        android:layout_height="wrap_content"
        android:orientation="vertical" >
        <ImageView
            android:layout_width="50dp"
            android:layout_height="50dp"
            android:layout_gravity="center"
            android:padding="5dp"
            android:src="@drawable/img_home2" />
        <TextView
            android:layout_width="match_parent"
            android:layout_height="wrap_content"
            android:gravity="center_horizontal"
            android:text="雀巢"
            android:textColor="#929292"
            android:textSize="13sp" />
    </LinearLayout>
    <LinearLayout
        android:id="@+id/ll_module3"
```

```xml
            android:layout_width="wrap_content"
            android:layout_height="wrap_content"
            android:orientation="vertical" >
            <ImageView
                android:layout_width="50dp"
                android:layout_height="50dp"
                android:layout_gravity="center"
                android:padding="5dp"
                android:src="@drawable/img_home3" />
            <TextView
                android:layout_width="match_parent"
                android:layout_height="wrap_content"
                android:gravity="center_horizontal"
                android:text="摩卡"
                android:textColor="#929292"
                android:textSize="13sp" />
        </LinearLayout>
        <LinearLayout
            android:id="@+id/ll_module4"
            android:layout_width="wrap_content"
            android:layout_height="wrap_content"
            android:orientation="vertical" >
            <ImageView
                android:layout_width="50dp"
                android:layout_height="50dp"
                android:layout_gravity="center"
                android:padding="5dp"
                android:src="@drawable/img_home4" />
            <TextView
                android:layout_width="match_parent"
                android:layout_height="wrap_content"
                android:gravity="center_horizontal"
                android:text="星巴克"
                android:textColor="#929292"
                android:textSize="13sp" />
        </LinearLayout>
    </TableRow>
    <TableRow
        android:layout_width="match_parent"
        android:layout_height="wrap_content"
        android:paddingBottom="6dp" >
        <LinearLayout
            android:id="@+id/ll_module5"
            android:layout_width="wrap_content"
```

```xml
            android:layout_height="wrap_content"
            android:orientation="vertical" >
            <ImageView
                android:layout_width="50dp"
                android:layout_height="50dp"
                android:layout_gravity="center"
                android:padding="5dp"
                android:src="@drawable/img_home5" />
            <TextView
                android:layout_width="match_parent"
                android:layout_height="wrap_content"
                android:gravity="center_horizontal"
                android:text="优惠券"
                android:textColor="#929292"
                android:textSize="13sp" />
        </LinearLayout>
        <LinearLayout
            android:id="@+id/ll_module6"
            android:layout_width="wrap_content"
            android:layout_height="wrap_content"
            android:orientation="vertical" >
            <ImageView
                android:layout_width="50dp"
                android:layout_height="50dp"
                android:layout_gravity="center"
                android:padding="5dp"
                android:src="@drawable/img_home6" />
            <TextView
                android:layout_width="match_parent"
                android:layout_height="wrap_content"
                android:gravity="center_horizontal"
                android:text="我的关注"
                android:textColor="#929292"
                android:textSize="13sp" />
        </LinearLayout>
        <LinearLayout
            android:id="@+id/ll_module7"
            android:layout_width="wrap_content"
            android:layout_height="wrap_content"
            android:orientation="vertical" >
            <ImageView
                android:layout_width="50dp"
                android:layout_height="50dp"
                android:layout_gravity="center"
```

```xml
            android:padding="5dp"
            android:src="@drawable/img_home7" />
        <TextView
            android:layout_width="match_parent"
            android:layout_height="wrap_content"
            android:gravity="center_horizontal"
            android:text="摇大奖"
            android:textColor="#929292"
            android:textSize="13sp" />
    </LinearLayout>
    <LinearLayout
        android:id="@+id/ll_module8"
        android:layout_width="wrap_content"
        android:layout_height="wrap_content"
        android:orientation="vertical" >
        <ImageView
            android:layout_width="50dp"
            android:layout_height="50dp"
            android:layout_gravity="center"
            android:padding="5dp"
            android:src="@drawable/img_home8" />
        <TextView
            android:layout_width="match_parent"
            android:layout_height="wrap_content"
            android:gravity="center_horizontal"
            android:text="物流查询"
            android:textColor="#929292"
            android:textSize="13sp" />
    </LinearLayout>
  </TableRow>
</TableLayout>
  <View
      android:layout_width="match_parent"
      android:layout_height="4dp"
      android:background="@color/form_background" />
  <LinearLayout
      android:id="@+id/recommand"
      android:layout_width="match_parent"
      android:layout_height="150dp"
      android:scrollbars="none"
      android:orientation="vertical"
      >
    <TextView
      android:layout_width="match_parent"
```

```xml
        android:layout_height="wrap_content"
        android:gravity="center_horizontal"
        android:text="推荐商品"
        android:textColor="#929292"
        android:textSize="15sp" />
    <GridView
        android:layout_width="match_parent"
        android:layout_height="wrap_content"
        android:numColumns="2"
        android:id="@+id/grid"/>"
</LinearLayout>
<View
    android:layout_width="match_parent"
    android:layout_height="4dp"
    android:background="@color/form_background" />
<LinearLayout
    android:id="@+id/discount"
    android:layout_width="match_parent"
    android:layout_height="150dp"
    android:background="#6666ff"
    android:orientation="vertical"
    >
<TextView
    android:layout_width="match_parent"
    android:layout_height="wrap_content"
    android:gravity="center_horizontal"
    android:text="打折商品"
    android:textColor="#929292"
    android:textSize="13sp" />
</LinearLayout>
<View
    android:layout_width="match_parent"
    android:layout_height="4dp"
    android:background="@color/form_background" />
<LinearLayout
    android:layout_width="match_parent"
    android:layout_height="150dp"
    android:scrollbars="none"
    android:background="#ff6666"
    android:orientation="vertical"
    >
<TextView
    android:layout_width="match_parent"
    android:layout_height="wrap_content"
```

```
            android:gravity="center_horizontal"
            android:text="热销商品"
            android:textColor="#929292"
            android:textSize="13sp" />
        </LinearLayout>
    </LinearLayout>
</ScrollView>
```

使用 ViewPager 控件可以实现左右滑动屏幕切换 View 的功能。这部分功能需要在 Activity 文件中编写代码来实现。要为 ViewPager 设置适配器的语句如下。

```
main_viewPager.setAdapter(new MyAdapter(getSupportFragmentManager(),
fragmentList));
```

详细的代码如下。

```
import android.os.Bundle;
import android.support.v4.app.Fragment;
import android.support.v4.app.FragmentActivity;
import android.support.v4.app.FragmentManager;
import android.support.v4.app.FragmentPagerAdapter;
import android.support.v4.view.ViewPager;
import android.support.v4.view.ViewPager.OnPageChangeListener;
import android.widget.RadioButton;
import android.widget.RadioGroup;
import android.widget.RadioGroup.OnCheckedChangeListener;
import com.example.jason.cofeestore.fragment.HomeFragment;
import com.example.jason.cofeestore.fragment.MeFragment;
import com.example.jason.cofeestore.fragment.SearchFragment;
import com.example.jason.cofeestore.fragment.ShopCarFragment;
import com.example.jason.cofeestore.fragment.SortFragment;
import java.util.ArrayList;
public class MainActivity extends FragmentActivity implements
OnCheckedChangeListener {
    //ViewPager
    private ViewPager main_viewPager;
    //RadioGroup
    private RadioGroup main_tab_RadioGroup;
    //RadioButton
    private RadioButton radio_home, radio_shopcar, radio_sort,
    radio_me, radio_search;
    private ArrayList<Fragment>fragmentList;
    @Override
    public void onCreate(Bundle savedInstanceState){
        super.onCreate(savedInstanceState);
        setContentView(R.layout.activity_main);
        //界面初始函数,用来获取定义的各控件对应的 ID
        InitView();
```

```java
        //ViewPager 初始化函数
        InitViewPager();
    }
    public void InitView(){
        main_tab_RadioGroup= (RadioGroup)findViewById(R.id.main_tab_RadioGroup);
        radio_home= (RadioButton)findViewById(R.id.radio_home);
        radio_shopcar= (RadioButton)findViewById(R.id.radio_shopcar);
        radio_sort= (RadioButton)findViewById(R.id.radio_sort);
        radio_search= (RadioButton)findViewById(R.id.radio_search);
        radio_me= (RadioButton)findViewById(R.id.radio_me);
        main_tab_RadioGroup.setOnCheckedChangeListener(this);
    }
    public void InitViewPager(){
        main_viewPager= (ViewPager)findViewById(R.id.main_ViewPager);
        fragmentList=new ArrayList<Fragment>();
        Fragment homeFragment=new HomeFragment();
        Fragment sortFragment=new SortFragment();
        Fragment shopCarFragment=new ShopCarFragment();
        Fragment searchFragment=new SearchFragment();
        Fragment meFragment=new MeFragment();
        fragmentList.add(homeFragment);
        fragmentList.add(shopCarFragment);
        fragmentList.add(sortFragment);
        fragmentList.add(searchFragment);
        fragmentList.add(meFragment);
        main_viewPager.setAdapter(new MyAdapter(getSupportFragmentManager(),
        fragmentList));
        main_viewPager.setCurrentItem(0);
        main_viewPager.addOnPageChangeListener(new MyListner());
    }
    public class MyAdapter extends FragmentPagerAdapter {
        ArrayList<Fragment>list;
        public MyAdapter(FragmentManager fm, ArrayList<Fragment>list){
            super(fm);
            this.list=list;
        }
        @Override
        public Fragment getItem(int arg0){
            return list.get(arg0);
        }
        @Override
        public int getCount(){
            return list.size();
        }
    }

    public class MyListner implements OnPageChangeListener {
```

```java
    @Override
    public void onPageScrollStateChanged(int arg0){

    }

    @Override
    public void onPageScrolled(int arg0, float arg1, int arg2){

    }

    @Override
    public void onPageSelected(int arg0){
        int current=main_viewPager.getCurrentItem();
        switch(current){
            case 0:
                main_tab_RadioGroup.check(R.id.radio_home);
                break;
            case 1:
                main_tab_RadioGroup.check(R.id.radio_shopcar);
                break;
            case 2:
                main_tab_RadioGroup.check(R.id.radio_sort);
                break;
            case 3:
                main_tab_RadioGroup.check(R.id.radio_search);
                break;
            case 4:
                main_tab_RadioGroup.check(R.id.radio_me);
                break;
        }
    }
}

@Override
public void onCheckedChanged(RadioGroup radioGroup, int checkId){
    int current=0;
    switch(checkId){
        case R.id.radio_home:
            current=0;
            break;
        case R.id.radio_shopcar:
            current=1;
            break;
        case R.id.radio_sort:
            current=2;
            break;
```

```
            case R.id.radio_search:
                current=3;
                break;
            case R.id.radio_me:
                current=4;
                break;
        }
        if(main_viewPager.getCurrentItem()!=current){
            main_viewPager.setCurrentItem(current);
        }
    }
}
```

7.6.3 项目说明

FragmentPagerAdapter 是 PagerAdapter 其中的一种实现。它将每一个页面表示为一个 Fragment,并且每一个 Fragment 都将会保存到 Fragment Manager 当中。当用户不可能再次回到页面的时候,Fragment Manager 才会将这个 Fragment 销毁。FragmentPagerAdapter 适用于一些静态 Fragment 时,使用时只需创建 FragmentPagerAdapter 的子类并且实现 getItem(int)和 getCount()方法即可。

在程序中添加了此部分项目实现中的代码后,可以实现如图 7.7 所示的运行结果,当滑动到首页或者单击首页按钮时,显示如图 7.7(a)所示的界面。当继续滑动或者单击不同的单选按钮,会切换到相应的 Fragment 页面,如图 7.7(b)所示。

(a)　　　　　　　　　　　　　　(b)

图 7.7　主页运行效果截图

本 章 小 结

本章主要介绍了 ViewPager 和 Fragment 的应用、ViewPager 结合 ViewPagerAdapter 等适配器的应用、Fragment 的应用场合和创建方法、Activity 和 Fragment 之间的调用以及它们生命周期的关系。还介绍了 Intent 的应用,最后介绍了 CoffeeStore 主页的实现,以上知识点都在此项目中得以应用。

本 章 习 题

1. 简述 Fragment 与 Activity 的交互方式。
2. 编写程序实现两个 Fragment 的切换。
3. 简述 Intent 对象是如何传递信息的。
4. 改写例 7.1,增加一个 Fragment,调试并运行程序。
5. 改写 CoffeeStore 项目,在项目增加一个 Fragment,表示订单查看界面,可以滑动,也可以通过选中单选按钮来控制选中。
6. 编写项目的分类界面,实现用 Intent 进行首页和分类页面的信息传递。

Android 高级控件

本章概述

通过本章的学习，掌握 Android 常用高级控件的用法，如 ListView、ViewFlipper、HorizontalScrollView 等。这些控件是 Android 开发中必不可少的，本章将对 Android 中的高级控件做全面清晰的讲解。

在本章的项目实现部分，可以使用这些高级控件来实现 CoffeeStore 店铺列表页的显示、首页广告条、启动页图片的切换以及首页分类条等功能模块。

学习重点与难点

重点：

（1）Adapter 与其他控件的适配。
（2）Spinner 的实现及显示优化。
（3）ListView 的实现方法与属性辨析。
（4）ExpandableListView 完成可扩展 ListView。
（5）GridView 实现九宫格效果。
（6）HorizontalScrollView 的使用。

难点：

（1）ListView 的实现与应用方法、性能优化。
（2）ExpandableListView 完成可扩展 ListView，并实现单击事件。

学习建议

本章需要掌握并熟练使用 ListView 控件，同时需要对 Adapter 内容有清晰准确的认识。很多控件，如 Spinner、ListView、HorizontalScrollView 都会与 Adapter 对象联合使用，所以学好使用 Adapter 对象非常关键。对于本章的扩展知识，如 GridView、ExpandableListView 等控件，建议多多熟悉并使用。

8.1 Adapter 对象

Adapter 称为适配器，是控件和数据源之间的桥梁，它可以将不同形式的数据源绑定到控件上。比如现实生活中的笔记本电源就叫电源适配器，它把 220V 的电源转化成电脑可以使用的 24V 电源。

本节所说的适配器,是为容器提供子视图,利用视图的数据和元数据来构建每个子视图。最终使数据和 UI(View)之间建立联系,所以常见控件用到 Adapter 的机会比较多。一般用到的数据库适配器是 CursorAdapter,数组对象的适配器是 ArrayAdapter,集合对象的适配器是 SimpleAdapter。下面重点介绍控件与 Adapter 的使用。

8.2 Spinner 控件

Spinner 控件实现类似下拉列表的功能,也将其称为下拉列表控件,单击该控件时会弹出一个列表选项。使用 Spinner 控件时,会创建 Adapter 对象,创建的过程中会指定要装载的数据。使用 Spinner 控件时的主要步骤如下。

(1) 获取 Spinner 对象。

(2) 创建 Adapter。

(3) 为 Spinner 对象设置 Adapter。

(4) 为 Spinner 对象设置监听器。

【例 8-1】 Spinner 控件示例。

下面编写如图 8.1 所示的 Spinner 控件测试用例。其中数据源来自字符串数组,主要功能是实现图 8.1(a)中的下拉列表,并使用单击事件,菜单被选中时显示图 8.1(b)所示的界面。

(a)

(b)

图 8.1 Spinner 控件测试用例

依照前文描述的主要步骤,该用例源代码文件如下。

(1) 首先获取 Spinner 对象。同其他控件一样,该 Spinner 对象需要在布局文件中定义,在 XML 文件中引入 Spinner 控件方法与其他控件的方法一样,这里不再重述。需要注意的是,本例是在 main.xml 文件中引入,并设置该控件 ID 为 mySpinner,下面的代码是在对应的 Java 文件中获取 Spinner 对象。

```java
public class SpinnerExample extends Activity {
private Spinner mySpinner;

/* 当 activity 第一次创建时调用 */
@Override
public void onCreate(Bundle savedInstanceState){
super.onCreate(savedInstanceState);
setContentView(R.layout.main);                    //创建 Spinner 对象
mySpinner=(Spinner)this.findViewById(R.id.mySpinner);
```

(2) 创建 Adapter。

这里的 R.array.citys 列表项是已经定义好的数组列表,具体使用方法可以参见本书值资源下数组资源的使用,数组的具体代码如下。

```xml
<string-array name="cities">
    <item>北京</item>
    <item>上海</item>
    <item>天津</item>
    <item>深圳</item>
</string-array>
```

另外,在开发过程中,需要对 Adapter 设置下拉列表样式(setDropDownViewResource),这里使用了编译环境里自带的样式,也可以选择其他下拉样式。

```java
ArrayAdapter<CharSequence>adapter=ArrayAdapter.createFromResource(
this, R.array.citys, android.R.layout.simple_spinner_item);
adapter.setDropDownViewResource (android.R.layout.simple_spinner_dropdown_item);
```

(3) 为 Spinner 对象设置 Adapter。

```java
mySpinner.setAdapter(adapter);
```

(4) 为 Spinner 对象设置监听器,此时,当条目被点击时传入 position、id 等参数,获取该参数,并通过 Toast 显示到屏幕中。

```java
mySpinner.setOnItemSelectedListener(new

OnItemSelectedListener(){
@Override
public void onItemSelected(AdapterView<?>parent, View view,
    int position, long id){
        Toast.makeText(
            SpinnerExample.this,
```

```
"position:"+position+" id:"+id+" value:"
+mySpinner.getSelectedItem().toString(),
Toast.LENGTH_SHORT).show();
}
```

为了响应单击事件,设置其单击事件 setOnItemSelectedListener,监听接口使用的方法是 OnItemSelectedListener void onItemSelected(AdapterView<?> parent, View view ,int position,long id)。其中第一个参数类似 context,它的知识范围较小,是指当前操作的 AdapterView,即父视图;第二个参数表示具体单击的那个 textView 对象,单击此 textView 对象在整个 AdapterView 中的位置,以及被单击 view 的 id。在程序开发过程中,需要动态修改 Spinner,即可以动态地增加或删除下拉列表,此时需要把静态数组改为 ArrayList,作为数据源。

8.3 ListView 控件

ListView 控件为列表视图,是以列表的形式显示数据。它能方便地列出一系列信息,如垂直重复信息。Android 程序中的许多列表式显示模式使用的都是 ListView 控件,它是 Android 开发中最常用的控件。

在 ListView 中,通过 Adapter 获得需要显示的数据。其中 SimpleAdapter 是扩展性非常好的适配器,SimpleAdapter 使用方便且可以定义各种想要的布局。SimpleAdapter 的参数功能如表 8.1 所示。

表 8.1 SimpleAdapter 的参数及功能介绍

参数名称	位置	参数功能及介绍
Context context	第一个参数	上下文,关联 SimpleAdapter 运行的视图上下文
List<? extends Map<String, ?>> data	第二个参数	数据源,需要的数据结构是一个 List<>里面的对象应该继承自 Map<String,?>
int resourceId	第三个参数	布局文件的 Id
String[]from	第四个参数	data 数据源中的"键",通过该键可以得到 data 中需要展示的 value
int[] to	第五个参数	item 布局文件中 Textview 的 Id,也就是数据要显示的具体位置

【例 8-2】 显示简单的 ListView 列表。

下面使用自定义布局建立如图 8.2 所示的单行自定义 ListView,加载已经写好的数组内容。

首先 Listview 创建并引用 layout,layout 里含有 ListView 想要展示的内容,控件的代码如下。

```
<ListView
    android:layout_width="wrap_content"
```

图 8.2 单行自定义 ListView

```
android:layout_height="wrap_content"
android:id="@+id/myListView"
android:layout_alignParentLeft="true"
android:layout_alignParentStart="true"/>
```

在 Java 文件中对该 ListView 加载具体内容,需显示的内容为下面定义的城市数组 cities,并且通过 Adapter 绑定 cities 与 ListView,代码如下。

```
//1.获取 ListView 对象,通过 findViewById 方法
myListView=(ListView)findViewById(R.id.myListView);
//2.定义一个数组,显示具体内容
String[] cities={"加格达奇","鄂尔多斯","呼伦贝尔"};
//3.写一个 adapter,不写具体内容,内容从之前定义好的资源文件里引用过来
ArrayAdapter<String>adapter=new ArrayAdapter<String>(this,
        R.layout.support_simple_spinner_dropdown_item,cities);
//4.创建 Adapter 对象,设置 Adapter
myListView.setAdapter(adapter);
```

这里介绍一下 R.layout.support_simple_spinner_dropdown_item 布局。该布局是显示 ListView 默认的布局,也可以对该布局的显示方式进行调整,使用自定义布局显示城市信息。上述代码只能选择一个条目,使用 setChoiceMode 实现对多条列表项的选择,代码如下。

```
listView.setChoiceMode(listView.CHOICE_MODE_MULTIPLE);
```

【例 8-3】 ListView 使用 SimpleAdapter 显示图文排列。

例 8-2 实现的是 ListView 简单文本显示效果,下面修改本例的 Java 代码,实现 ListView 图文显示效果。首先在主布局文件 activity_main.xml 中定义 ListView,代码如下。

```
<ListView
    android:layout_width="match_parent"
    android:layout_height="wrap_content"
    android:divider="#5be60a"
    android:dividerHeight="10sp"
    android:listSelector="#f27611"
    android:id="@+id/listView1"
```

```
></ListView>
```

为了使 List 显示图文效果，使用一个自定义的布局文件 activity_main2.xml，它的主要作用是定义 List 需要显示的图文，关键代码如下。

```
<ImageView
    android:layout_width="wrap_content"
    android:layout_height="wrap_content"
    android:src="@mipmap/ic_launcher"
    />
<TextView
    android:layout_width="wrap_content"
    android:layout_height="wrap_content"
    android:textSize="30sp"
    android:text="Hello world"
    android:id="@+id/textView_icon"/>
```

在 Activity 中首先获取 ListView，代码如下。

```
setContentView(R.layout.activity_main);
listView1=(ListView)findViewById(R.id.listView1);
```

接下来定义一个 list，并在 list 里添加具体内容，在 SimpleAdapter 中使用该 list 中的内容，关键代码如下。

```
//准备数据，每一个 HashMap 就是一条记录
HashMap<String, String>title1=new HashMap<>();
title1.put("title","这是第 1 行 HashMap");
HashMap<String, String>title2=new HashMap<>();
title2.put("title","文字可以随意填写");
HashMap<String, String>title3=new HashMap<>();
title3.put("title","通过 HashMap 加入 list 列表中");
//建立 List 列表
List<Map<String,String>>list=new ArrayList<>();
//将 HashMap 添加到 list 中
list.add(title1);
list.add(title2);
list.add(title3);
//1.上下文 2.List 列表，内部是 map 集合 3.填充到布局 4.数据 5.控件
SimpleAdapter sa = new SimpleAdapter (this, list, R.layout.activity_main2, new String[] {"title"},new int[]{R.id.textView_icon});
listView1.setAdapter(sa);
```

显示效果如图 8.3 所示。

前面介绍了如何在程序中显示不同样式的列表，下面实现当发生单击、滑动等事件时的处理，即 ListView 事件的处理。

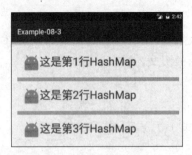

图 8.3　显示图文排列

ListView 事件的处理会使用 BaseAdapter，BaseAdapter 是各种 Adapter 的父类，它支持自定义的数据绑定与显示。自定义的 Adapter 继承 BaseAdapter 时，需要重写父类的以下几个方法。

- getCount():int
- getItem(position:int):Object
- getItemId(position:int):int
- getView(position:int, convertView:View, parent:ViewGroup):View

【例 8-4】　商品列表页实现。

下面使用 ListView 控件实现更加复杂的商品列表页，实现该功能的主要步骤如下。

(1) 首先创建并加载布局资源 activity_main，该布局中只定义了一个 ListView 控件，设置了单击时的颜色属性 listSelector。

```xml
<?xml version="1.0" encoding="utf-8"?>
<LinearLayout xmlns:android="http://schemas.android.com/apk/res/android"
    xmlns:tools="http://schemas.android.com/tools"
    android:id="@+id/activity_main3"
    android:layout_width="match_parent"
    android:layout_height="match_parent"
    android:paddingBottom="@dimen/activity_vertical_margin"
    android:paddingLeft="@dimen/activity_horizontal_margin"
    android:paddingRight="@dimen/activity_horizontal_margin"
    android:paddingTop="@dimen/activity_vertical_margin"
    android:orientation="vertical"
    tools:context="cn.edu.neusoft.spinner.Main3Activity">
    <ListView
        android:layout_width="wrap_content"
        android:layout_height="wrap_content"
        android:divider="@color/colorPrimary"
        android:dividerHeight="1sp"
        android:listSelector="#ffe100"
        android:id="@+id/listView1"
        >
    </ListView>
```

```
</LinearLayout>
```

（2）创建布局参数，设置 ListView 中每一个条目的显示样式，该布局文件名字定义为 list_main。

```
<?xml version="1.0" encoding="utf-8"?>
<LinearLayout xmlns:android="http://schemas.android.com/apk/res/android"
    xmlns:tools="http://schemas.android.com/tools"
    android:id="@+id/activity_main"
    android:layout_width="match_parent"
    android:layout_height="match_parent"
    android:paddingBottom="@dimen/activity_vertical_margin"
    android:paddingLeft="@dimen/activity_horizontal_margin"
    android:paddingRight="@dimen/activity_horizontal_margin"
    android:paddingTop="@dimen/activity_vertical_margin"
    android:orientation="horizontal"
    tools:context="cn.edu.neusoft.spinner.MainActivity">
    <ImageView
        android:layout_width="80sp"
        android:layout_height="80sp"
        android:scaleType="centerInside"
        android:id="@+id/img"
        android:src="@mipmap/ic_launcher"/>
    <LinearLayout
        android:layout_width="wrap_content"
        android:layout_height="wrap_content"
        android:layout_marginLeft="10dp"
        android:orientation="vertical">
    <TextView
        android:layout_width="wrap_content"
        android:layout_height="wrap_content"
        android:textSize="15sp"
        android:textColor="#131111"
        android:id="@+id/tv"
        android:text="Hello World!" />
    <TextView
        android:layout_width="wrap_content"
        android:layout_height="wrap_content"
        android:id="@+id/price"
        android:text="Hello World!" />
        <TextView
            android:layout_width="wrap_content"
            android:layout_height="wrap_content"
            android:id="@+id/discount"
            android:background="#ff5647"
```

```
                android:textColor="#fff"
                android:text="Hello World!" />
        </LinearLayout>
</LinearLayout>
```

（3）在 Activity 中获取列表项显示的数据，加载需要显示的信息并显示到屏幕上，这里使用的是 Map<String，Object>。逐一添加细节信息，最后设置列表项的单击事件，代码如下。

```
public class Main3Activity extends AppCompatActivity {
    private ListView listView1;
    @Override
    protected void onCreate(Bundle savedInstanceState){
        super.onCreate(savedInstanceState);
        setContentView(R.layout.activity_main3);
        listView1=(ListView)findViewById(R.id.listView1);
        HashMap<String,Object>title1=new HashMap<>();
        title1.put("img", R.drawable.img1);
        title1.put("title", "荣耀 畅玩 6X 4GB 32GB 全网通 4G 手机 ");
        title1.put("price", "￥1299.00");
        title1.put("discount", "9折");
        HashMap<String,Object>title2=new HashMap<>();
        title2.put("img", R.drawable.img2);
        title2.put("title", "小米 红米 4A 全网通 2GB 内存 16GB ROM 香槟金色");
        title2.put("price", "￥599.00");
        title2.put("discount", "8.5折");
        HashMap<String,Object>title3=new HashMap<>();
        title3.put("img", R.drawable.img3);
        title3.put("title", "诺基亚 6(Nokia6)4GB+ 64GB 黑色 全网通");
        title3.put("price", "￥1699.00");
        title3.put("discount", "9折");
        HashMap<String,Object>title4=new HashMap<>();
        title4.put("img", R.drawable.img4);
        title4.put("title", "OPPO A57 3GB+ 32GB 内存版 ");
        title4.put("price", "￥1599.00");
        title4.put("discount", "9折");
        final List<HashMap<String,Object>>list=new ArrayList<>();
        list.add(title1);
        list.add(title2);
        list.add(title3);
        list.add(title4);
        SimpleAdapter sa=new SimpleAdapter(this,list,R.layout.list_main,new
        String[]{"img","title","price","discount"},new int[]{R.id.img,R.id.tv,
        R.id.price,R.id.discount});
        listView1.setAdapter(sa);
```

```
listView1.setOnItemClickListener(new AdapterView.OnItemClickListener(){
    @Override
    public void onItemClick(AdapterView<?>adapterView, View view, int i,
    long l){
        Toast.makeText(Main3Activity.this,"您点击的是"+list.get(i).get
        ("title").toString(),Toast.LENGTH_SHORT).show();
    }
});
}
```

在 Genymotion 模拟器中单击 ListView 的列表项时,会弹出对应的 title 文字,该商品列表页的实现如图 8.4 所示。

【例 8-5】 实现图 8.5 所示的书籍评分界面。主要通过 ListView 控件实现书籍名称、作者、价格、评分等界面显示效果。

图 8.4 商品列表页的实现

图 8.5 书籍评分界面

其中布局文件使用线性布局,排列方式为 horizontal,并且该布局中使用 ImageView 控件作为其他部分的背景图片。在这个布局基础上需要嵌套新的线性布局,新布局以 vertical 方式排列,这样就可以在原来布局的基础上横向排列新的控件。另外为了实现评分效果,该页面也采用了 RatingBar 控件。

XML 代码如下。

```
<?xmlversion="1.0"encoding="utf-8"?>
<LinearLayoutxmlns:android="http://schemas.android.com/apk/res/android"
    android:layout_width="match_parent"
    android:layout_height="wrap_content"
```

```xml
        android:orientation="horizontal">
<ImageView
android:id="@+id/img"
android:layout_width="60px"
android:layout_height="90px"
android:layout_marginTop="0px"
    />
<LinearLayout
android:layout_width="140px"
android:layout_height="90px"
android:orientation="vertical"
android:baselineAligned="@id/img"
android:layout_marginLeft="10px">
<TextView
    android:id="@+id/bookTitle"
    android:layout_width="wrap_content"
    android:textSize="8sp"
    android:layout_height="wrap_content"
    />
<TextView
    android:id="@+id/bookAuthor"
    android:layout_width="wrap_content"
    android:layout_height="wrap_content"
    android:textSize="8sp"
    />
<TextView
    android:id="@+id/bookPrice"
    android:layout_width="wrap_content"
    android:layout_height="wrap_content"
    android:textSize="8sp"
    android:textColor="#f00"
    />
<RatingBar
    android:id="@+id/ratingBar1"
    android:layout_width="wrap_content"
    android:layout_height="wrap_content"
    android:textSize="24sp"
    style="?android:attr/ratingBarStyleSmall"/>"
</LinearLayout>
    </LinearLayout>
```

Activity 及执行结果代码如下。

```java
public class MainActivity extends Activity{
private ListView listView;
```

```java
private ArrayList<HashMap<String, Object>>bookSource;
publicvoid getSource ( intimgId, String bookName, String bookAuthor,
intbookPrice){
        HashMap<String, Object>map=new HashMap<String, Object>();
map.put("img",imgId);
map.put("bookName",bookName);
map.put("bookAuthor",bookAuthor);
map.put("bookPrice",bookPrice);
bookSource.add(map);
}
@Override
protectedvoid onCreate(Bundle savedInstanceState){
super.onCreate(savedInstanceState);
setContentView(R.layout.activity_main);
listView= (ListView)this.findViewById(R.id.listView1);
bookSource=new ArrayList<HashMap<String, Object>>();
getSource(R.drawable.img1,"疯狂 android 讲义","李刚",74);
getSource(R.drawable.img2,"深入理解 android","邓凡平",54);
getSource(R.drawable.img3,"android 内核剖析","柯元",64);
getSource(R.drawable.img4,"android 高级编程","李五",44);
getSource(R.drawable.img5,"android 开发入门与实战","李无",55);
getSource(R.drawable.img6,"android 开发入门与实战第二版","王艺",88);
getSource(R.drawable.img7,"android 4 高级编程","李我",66);
getSource(R.drawable.img9,"android 开发高清在线视频","李兴化",99);
SimpleAdapter adapter=new MyAdapter
        (this,bookSource,R.layout.list_item_custom,
         new String[]{"img","bookName","bookAuthor","bookPrice"},
newint[]{R.id.img,R.id.bookTitle,R.id.bookAuthor,R.id.bookPrice});
listView.setAdapter(adapter);
}
class MyAdapter extends SimpleAdapter {
public MyAdapter(Contextcontext, List<?extends Map<String, ?>>data,
intresource, String[] from, int[] to){
super(context, data, resource, from, to);
//TODO Auto-generated constructor stub
}
@Override
public View getView(intposition, View convertView, ViewGroup parent){
//TODO Auto-generated method stub
View view=super.getView(position, convertView, parent);
TextView txt= (TextView)view.findViewById(R.id.bookTitle);
        if(position %2 ==0){
            view.setBackgroundColor(Color.GRAY);
txt.setTextColor(Color.RED);
```

```
        } else {
                view.setBackgroundColor(Color.WHITE);
    txt.setTextColor(Color.GREEN);
        }
            returnview;
        }
        }
    }
```

上例使用 SimpleAdapter 实现了 ListView 控件中图文排列的效果,但是对一些特殊的控件(比如 Button)来说,单击 ListView 控件会产生一些混乱。当开发者在 ListView 的条目中引用一个 Button 控件,用户单击该控件时就会造成混乱,程序可能无法识别用户是单击 ListView 的条目还是 Button 控件。为了避免这种情况发生,需要使用 BaseAdapter 来替代 SimpleAdapter。

BaseAdapter 是各种 Adapter 的父类,它支持自定义的数据绑定与显示,使用 BaseAdapter 时需要重写以下几个方法。

- getCount()
- getItem(int position)
- getItemId(int position)
- getView(int position,View convertView,ViewGroup parent)

【例 8-6】 使用 BaseAdapter 实现书籍评分案例。

下面对例 8-5 的知识进行扩展,通过使用 BaseAdapter 实现书籍显示案例。首先在主布局文件中添加一个 ListView,将该 ListView 命名为 listView,代码如下。

```
<ListView
    android:layout_width="match_parent"
    android:layout_height="match_parent"
    android:id="@+id/listView"
    android:layout_alignParentLeft="true"
    android:layout_alignParentStart="true"
    android:layout_alignParentTop="true">
</ListView>
```

同时定义 ListView 控件每个条目的显示样式,本例定义了一个名为 list_item 的布局文件,内部定义了一个 ImageView 控件和一个 TextView 控件。代码如下。

```
<?xml version="1.0" encoding="utf-8"?>
<LinearLayout xmlns:android="http://schemas.android.com/apk/res/android"
    android:layout_width="match_parent"
    android:layout_height="match_parent"
    android:orientation="horizontal">

    <ImageView
        android:id="@+id/imageView_icon"
```

```xml
        android:layout_width="90dp"
        android:layout_height="120dp" />

    <TextView
        android:id="@+id/textView_title"
        android:layout_width="wrap_content"
        android:layout_height="wrap_content"
        android:layout_margin="10dp" />
</LinearLayout>
```

定义 ImageView 控件的同时,将其需要引用的图片放置在 drawable 文件夹中,命名为 img1 到 img5。此时已经将布局文件和资源文件放置在开发环境中,以下代码使用了 BaseAdapter 完成书籍显示案例。

```java
public class MainActivity extends AppCompatActivity {
    private ListView listView;

    @Override
    protected void onCreate(Bundle savedInstanceState) {
        super.onCreate(savedInstanceState);
        setContentView(R.layout.activity_main);
        listView=(ListView)findViewById(R.id.listView);
        listView.setAdapter(new MyAdapter(this));
    }

    static class MyAdapter extends BaseAdapter {
        private String[] titles={"书籍1","书籍2","书籍3","书籍4","书籍5"};
        private int[] icons={R.drawable.img1,
                R.drawable.img2, R.drawable.img3,
                R.drawable.img4, R.drawable.img5,};
        private Context context;

        //构造方法传进来一个 context,用它来实例化 layoutinflatter
        public MyAdapter(Context context){
            this.context=context;
        }

        @Override
        public int getCount(){
            return titles.length;
        }
        @Override
        public Object getItem(int position){
            return titles[position];
        }
```

```
        @Override
        public long getItemId(int position){
            return position;
        }
        @Override
        public View getView(int position, View convertView, ViewGroup parent){
            LayoutInflater inflater=LayoutInflater.from(context);
            //实例化布局文件
            View v=inflater.inflate(R.layout.list_item, null);
            TextView tv_title= (TextView)v.findViewById(R.id.textView_title);
            ImageView iv_icon= (ImageView)v.findViewById(R.id.imageView_icon);
            tv_title.setText(titles[position]);
            iv_icon.setImageResource(icons[position]);
            return v;
        }
    }
}
```

上述代码对 getCount()、getItem()、getItemId()、getView()方法进行了重写,其中在 getView()方法中使用了 inflater.inflate()方法,引入了自定义布局 list_item,从而实现了 ListView 条目中各个控件的使用。

【例 8-7】 优化 ListView 性能。

在程序开发中,ListView 控件的使用十分频繁,如果能优化 ListView 控件性能就会大大提高程序的可用性。下面对例 8-6 的知识进行扩展,实现性能优化。定义书籍标题和图片时,增加 10 项内容,使程序运行时 ListView 内容布满整个屏幕。代码修改如下。

```
private String[] titles={"书籍 1", "书籍 2", "书籍 3", "书籍 4", "书籍 5", "书籍 6", "书籍 7", "书籍 8", "书籍 9", "书籍 10", "书籍 11", "书籍 12", "书籍 13","书籍 14", "书籍 15"};
private int[] icons={R.drawable.img1,
        R.drawable.img2, R.drawable.img3,
        R.drawable.img4, R.drawable.img5,
        R.drawable.img1, R.drawable.img2,
        R.drawable.img3, R.drawable.img4,
        R.drawable.img5, R.drawable.img1,
        R.drawable.img2, R.drawable.img3,
        R.drawable.img4, R.drawable.img5};
```

程序运行时,由于 ListView 内容布满整个屏幕,向上或向下拖曳时,一个新的条目产生程序会创建一个内存空间给该条目。多次操作该 ListView 控件程序会持续创建内存空间,这就造成程序会占用越来越多的资源。为便于说明,在 getView()方法中添加如下代码进行监控。

```
System.out.println(position+"---此时创建新的内存空间:---"+convertView);
```

在模拟器中,当用户滑动 ListView 控件时,开发环境 logcat 打印语句如下。

9---此时创建新的内存空间:---android.widget.LinearLayout{f08163 V.E..............0,-166-1136,74}
10---此时创建新的内存空间:---android.widget.LinearLayout{afcb619 V.E............0,56-1136,296}
11---此时创建新的内存空间:---android.widget.LinearLayout{afcb619 V.E............0,56-1136,296}
12---此时创建新的内存空间:---android.widget.LinearLayout{7d202ea V.E..............0,-163-1136,77}
13---此时创建新的内存空间:---android.widget.LinearLayout{ad40c78V.E..............0,189-1136,429}
14---此时创建新的内存空间:---android.widget.LinearLayout{ad40c78V.E..............0,189-1136,429}
7---此时创建新的内存空间:---android.widget.LinearLayout{84563b6 V.E......0,-53-1136,187}
6--此时创建新的内存空间:--android.widget.LinearLayout{eb118b7V.E............0,1582-1136,1822}

从以上语句看到程序在不断创建内存空间,此时可以使用 getView()方法中的 convertView 参数实现内存空间循环使用。该参数为空闲可用对象,当 ListView 控件条目被滑出屏幕时,就会产生空闲可用对象,可以使用该参数实现内存重复使用。优化 ListView 的加载速度,就要让 convertView 匹配列表类型,并最大程度地重新使用 convertView。

具体代码如下。

```
@Override
public View getView(int position, View convertView, ViewGroup parent){
    if(convertView ==null){
        LayoutInflater inflater=LayoutInflater.from(context);
        //实例化布局文件
        convertView=inflater.inflate(R.layout.list_item, null);
    }
    System.out.println(position+"---此时创建新的内存空间:---"+convertView);
    TextView tv_title= (TextView)convertView.findViewById(R.id.textView_title);
    ImageView iv_icon= (ImageView)convertView.findViewById(R.id.imageView_icon);
    tv_title.setText(titles[position]);
    iv_icon.setImageResource(icons[position]);
    return convertView;
}
```

上述代码的作用是当 convertView 不为空时直接重新使用 convertView,从而减少了很多不必要的 View 的创建。程序运行时,滑动控件 logcat,打印语句如下,其中条目 1 使用了 9 的内存空间,0 使用了 10 的内存空间。

1---此时创建新的内存空间:---android.widget.LinearLayout{877d5cf V.E...... . 0,1259-1136,1499}
0---此时创建新的内存空间:---android.widget.LinearLayout{8b50a9 V.E...... .. 0,1501-1136,1741}
5---此时创建新的内存空间:---android.widget.LinearLayout{33b6be1 V.E...... 0,973-1136,1213}
6---此时创建新的内存空间:---android.widget.LinearLayout{d21e51d V.E...... 0,1215-1136,1455}
7---此时创建新的内存空间:---android.widget.LinearLayout{ba0438cV.E...... 0,1457-1136,1697}
8---此时创建新的内存空间:---android.widget.LinearLayout{e9f4e65 V.E...... 0,256-1136,496}
9---此时创建新的内存空间:---android.widget.LinearLayout{877d5cf V.E...... 0,14-1136,254}
10---.-此时创建新的内存空间:---android.widget.LinearLayout{8b50a9 V.E...... 0,-228-1136,12}

【例 8-8】 优化 ListView 性能 2。

为优化 ListView 性能,可以使用 ViewHolder,它是一个内部类,其中包含了单个项目布局中的各个控件。由于程序中经常使用 findViewById() 方法查找空间,而该方法的实现需要对每个 id 进行匹配,这就会造成资源的浪费。ViewHolder 用于保存第一次查找的组件,这就可以避免下次重复查找,从而节约系统资源。

使用 ViewHolder 对例 8-7 的 getView() 方法进行优化,代码如下。

```
@Override
    public View getView(int position, View convertView, ViewGroup parent){
        ViewHolder vh;
        if(convertView ==null){
            LayoutInflater inflater=LayoutInflater.from(context);
            convertView=inflater.inflate(R.layout.list_item, null);
            vh=new ViewHolder();
            vh.iv_icon=(ImageView)convertView.findViewById(R.id.imageView_icon);
            vh.tv_title=(TextView)convertView.findViewById(R.id.textView_title);
            convertView.setTag(vh);
        } else {
            vh=(ViewHolder)convertView.getTag();
        }
        vh.tv_title.setText(titles[position]);
        vh.iv_icon.setImageResource(icons[position]);
        return convertView;
    }

    static class ViewHolder {
        ImageView iv_icon;
```

```java
        TextView tv_title;
    }
}
```

程序最终代码如下。

```java
public class  MainActivity extends AppCompatActivity {
    private ListView listView;

    @Override
    protected void onCreate(Bundle savedInstanceState){
        super.onCreate(savedInstanceState);
        setContentView(R.layout.activity_main);
        listView= (ListView) findViewById(R.id.listView);
        listView.setAdapter(new MyAdapter(this));
    }

    static class MyAdapter extends BaseAdapter {
        private String[] titles={"书籍 1","书籍 2","书籍 3","书籍 4","书籍 5","书籍 6","书籍 7","书籍 8","书籍 9","书籍 10","书籍 11","书籍 12","书籍 13","书籍 14","书籍 15"};
        private int[] icons={R.drawable.img1,
                R.drawable.img2, R.drawable.img3,
                R.drawable.img4, R.drawable.img5,
                R.drawable.img1, R.drawable.img2,
                R.drawable.img3, R.drawable.img4,
                R.drawable.img5, R.drawable.img1,
                R.drawable.img2, R.drawable.img3,
                R.drawable.img4, R.drawable.img5};
        private Context context;

        //构造方法传进来一个 context,用它来实例化 layoutinflatter
        public MyAdapter(Context context){
            this.context=context;
        }

        @Override
        public int getCount(){
            return titles.length;
        }

        @Override
        public Object getItem(int position){
            return titles[position];
        }
```

```
@Override
public long getItemId(int position){
    return position;
}

@Override
//convertView 空闲可用对象
public View getView(int position, View convertView, ViewGroup parent){
    ViewHolder vh;
    if(convertView ==null){
        LayoutInflater inflater=LayoutInflater.from(context);
        //实例化布局文件
        convertView=inflater.inflate(R.layout.list_item, null);
        vh=new ViewHolder();
        vh.iv_icon= (ImageView)convertView.findViewById(R.id.imageView_icon);
        vh.tv_title= (TextView)convertView.findViewById(R.id.textView_title);
        convertView.setTag(vh);
    }
    else {
        vh= (ViewHolder)convertView.getTag();
    }
    vh.tv_title.setText(titles[position]);
    vh.iv_icon.setImageResource(icons[position]);
    //          tv_title.setText(titles[position]);
    //          iv_icon.setImageResource(icons[position]);
    return convertView;

}
static class ViewHolder {
    ImageView iv_icon;
    TextView tv_title;
}
    }
}
```

8.4　ExpandableListView 控件

ExpandableListView 控件是一个可以展开的列表控件，形式与菜单相似。控件未单击时正常显示，组列表项被单击时就会弹出子菜单，显示当前列表项下的所有子列表项，再次单击该列表项会返回原界面。

ExpandableListView 控件对数据进行一个简单的罗列，只是一个简单的 ListView 的使用。若想实现类似于腾讯聊天工具 QQ 的效果，首先显示分组名称列表，接着单击某个组名选项,会弹出该组里的人员信息列表,在实际应用中,ExpandableListView 控件只允许两个层次。在使用 ExpandableListView 控件时,可扩展列表包含两层,一层是 GroupLayout,另一层是 ChildLayout。

ExpandableListView 控件是 ListView 的子类,所以拥有 ListView 的特性。为了设置和使用数据,使用 ExpandableListView 控件时也需要一个 Adapter 对象,此处的适配器类为 ExpandableAdapter。通过导入 LayoutManager 控制显示方式,ItemDecoration 控制 Item 间的间隔 ItemAnimator 控制动画,控制单击、长按事件需要开发者重写。

【例 8-9】 ExpandableListView 示例。

首先定义 ExpandableListView 布局,这里布局文件与 ListView 控件相似,这里定义了该控件的 id 为 expandableListView。

```
<ExpandableListView
    android:layout_width="match_parent"
    android:layout_height="match_parent"
    android:id="@+id/expandableListView"
    android:layout_alignParentTop="true"
    android:layout_alignParentLeft="true"
    android:layout_alignParentStart="true">
</ExpandableListView>
```

定义完该布局文件后,需要定义可扩展 ListView 的父元素与子元素的布局文件,分别为 children_layout.xml 和 group_layout.xml 文件,其中 group 布局和 children 类似,关键代码如下。

```
<ImageView
    android:layout_width="wrap_content"
    android:layout_height="wrap_content"
    android:id="@+id/icon_group"
    android:src="@drawable/ic_launcher"/>
<TextView
    android:layout_width="wrap_content"
    android:layout_height="wrap_content"
    android:id="@+id/title_group"
    android:textSize="30sp"
    android:text="hello"/>
```

在 Activity 中,首先获取 activity_main 文件中的 expandableListView。同时需要定义扩展子项的数据,定义了 children 这个子项后,单击好友栏时可以显示该子项的"蛋蛋""小 A""大哥"等数据,单击班级时可显示"AA""BB"等子项。

```
private String[] groups={"好友","班级"};
private String[][] children={{"蛋蛋","小 A","大哥"},{"AA","BB"}};
class MyExPandableAdapter extends BaseExpandableListAdapter{
```

```java
@Override
public boolean isChildSelectable(int groupPosition, int childPosition){
    return true;
}

@Override
public int getGroupCount(){
    return groups.length;
}

@Override
public int getChildrenCount(int groupPosition){
    return children[groupPosition].length;
}
@Override
public Object getGroup(int groupPosition){
    return groups[groupPosition];
}
@Override
public Object getChild(int groupPosition, int childPosition){
    return children[groupPosition][childPosition];
}
@Override
public long getGroupId(int groupPosition){
    return groupPosition;
}
@Override
public long getChildId(int groupPosition, int childPosition){
    return childPosition;
}
@Override
public boolean hasStableIds(){
    return false;
}
@Override
public View getGroupView (int groupPosition, boolean isExpanded, View convertView, ViewGroup parent){
    if(convertView ==null){
        convertView=getLayoutInflater().inflate(R.layout.group_layout, null);
    }
   ImageView icon= (ImageView)convertView.findViewById(R.id.icon_group);
    TextView title= (TextView)convertView.findViewById(title_group);
    title.setText(groups[groupPosition]);
    return convertView;
}
@Override
```

```
    public View getChildView (int groupPosition, int childPosition, boolean
isLastChild, View convertView, ViewGroup parent){
        if(convertView ==null){
            convertView=getLayoutInflater().inflate(R.layout.children_layout, null);
        }
        ImageView icon= (ImageView)convertView.findViewById(R.id.icon_child);
        TextView title= (TextView)convertView.findViewById(title_child);
        title.setText(children[groupPosition][childPosition]);
        return convertView;
    }
}
```

上述代码定义了一个 MyExPandableAdapter，并导入了各种方法，其中 getGroupView 方法实现了父级组的设置，而 getChildView 方法实现了子项的设置，这里也使用了 convertView 进行性能上的优化。设置完成后对该 listView 设置 Adapter，并对列表项设置单击方式，代码如下。

```
listView.setAdapter(new MyExPandableAdapter());
listView.setOnChildClickListener(new ExpandableListView.OnChildClickListener
(){
    @Override
    public boolean onChildClick (ExpandableListView parent, View v, int
    groupPosition, int childPosition, long id){

Toast.makeText (MainActivity.this, children [groupPosition] [childPosition],
Toast.LENGTH_SHORT).show();
        return true;
    }
});
```

最终显示结果如图 8.6 所示。单击好友时，会弹出"蛋蛋""小 A""大哥"等可扩展子项。

图 8.6　ExpandableListView 控件显示结果

8.5　GridView 控件

GridView 控件主要通过表格形式展示数据，通常用于定义类似"九宫格""十六宫格"等样式的布局。前面学习了 ListView，如果 Activity 中显示的是列表（单列多行形式），建议使用 ListView，如果是多行多列网状形式，则优先使用 GridView。

GridView（网格视图）是按照行列的方式显示内容的，一般用于显示图片、图片等内容，比如实现九宫格图，GridView 是首选。图 8.7 就是使用 GridView 控件实现的九宫格样式。

图 8.7　GridView 控件实现的
九宫格样式

【例 8-10】 GridView 控件示例。

首先定义布局文件样式，GridView 控件与其他控件的设置样式相似，具有控件的基本属性。这里重点说明一下 numColumns 属性，可以设置为具体值，比如以下代码设置了 3 列排序，也可以设置为 auto_fit，这时 Android 就会自动计算手机屏幕的大小，以决定每一行展示几个元素。

```
<GridView
android:id="@+id/gridview1"
android:layout_width="fill_parent"
android:layout_height="fill_parent"
android:numColumns="3"
android:horizontalSpacing="15px"
android:verticalSpacing="15px">
</GridView>
```

以上属性中 android：horizontalSpacing 定义的是列之间的间隔，android：verticalSpacing 定义行之间的间隔。另外，也可以将 android：stretchMode 设置为 columnWidth，意味着根据列宽自动缩放。

```
publicclassGridViewDemoActivityextends Activity{
private GridView gridView;
@Override
protectedvoid onCreate(Bundle savedInstanceState){
super.onCreate(savedInstanceState);
setContentView(R.layout.gridview_demo_layout);
gridView= (GridView)this.findViewById(R.id.gridView1);
gridView.setAdapter(new GridViewDataAdapter(this));
```

上述代码是在 Activity 文件中首先定义并获取该控件，并为该控件设置 Adapter，这里重写了一个 GridViewDataAdapter。该 Adapter 中定义了 GridView 的数据源，需要使用图片资源中的不同图片，使用这些图片之前需要定义数组 items，该数组存放的是 GridView 中需要显示的图片资源文件。

```
public class GridViewAdapter extends BaseAdapter {
    privatefinal Context context;
    privateint[] items;
    public GridViewDataAdapter(Context context){
this.context=context;
        this.fillitems();}
    Private void fillitems(){
items=new int[9];
items[0]=R.drawable.b_001;
items[1]=R.drawable.b_002;
items[2]=R.drawable.b_003;
items[3]=R.drawable.b_004;
items[4]=R.drawable.b_005;
items[5]=R.drawable.b_006;
items[6]=R.drawable.b_007;
items[7]=R.drawable.b_008;
items[8]=R.drawable.b_009;
}
@Override
publicint getCount(){
        returnitems.length;
}
@Override
public Object getItem(intposition){
        returnitems[position];
}
@Override
publiclong getItemId(intposition){
        returnposition;
}
@Override
public View getView(intposition, View convertView, ViewGroup parent){
if(convertView ==null){
            convertView=new ImageView(context);
}
        ImageViewimg= (ImageView)convertView;
        img.setImageResource(items[position]);
        return img;
}
```

练习：用九宫格的方式显示资源中的所有图片，并以九宫格的形式显示所有启动项，此时需要自定义 Adapter，最终显示效果如图 8.8 所示。

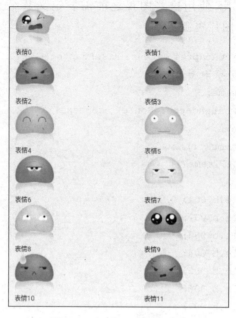

图 8.8　Gridview 控件显示

8.6　HorizontalScrollView 控件

移动设备物理显示空间一般有限，不可能一次性把所有要显示的内容都显示在屏幕上。所以各大平台一般会提供一些可滚动的视图来展示数据。HorizontalScrollView 控件就是可以实现一组图片轮播的效果，它是 FrameLayout 的子类，在它下面只能放置一个子控件，该子控件可以包含很多数据内容。这个子控件本身就可能是一个布局控件，可以包含非常多的其他展示数据的控件。这个布局控件一般使用的是一个水平布局的 LinearLayout。

下面介绍一个 HorizontalScrollView 中包含许多图片，并且可以滚动浏览的示例。布局方式使用的是 LinearLayout，方向设置为水平，具体代码如下。

【例 8-11】　HorizontalScrollView 实例。

```
<LinearLayout xmlns:android="http://schemas.android.com/apk/res/android"
    xmlns:tools="http://schemas.android.com/tools" android:layout_width=
    "match_parent"
    android:layout_height="match_parent">
    <HorizontalScrollView android:layout_width="wrap_content" android:layout_
    height="150dp"
        android:layout_gravity="center_vertical" android:background=
        "#AA444444"
        android:scrollbars="none">
        <LinearLayout android:id="@+id/id_gallery" android:layout_width="wrap
```

```
        _content"
        android:layout_height="wrap_content" android:layout_gravity=
        "center_vertical"
        android:orientation="horizontal"></LinearLayout>
    </HorizontalScrollView>

</LinearLayout>
```

之后在 MainActivity 中引用该布局文件,并且将 drawable 下的图片文件循环加入 HorizontalScrollView 的 LinearLayout 中,具体代码如下。

```
public class MainActivity extends Activity {
    private LinearLayout mGallery;
    private int[] mImgIds;
    private LayoutInflater mInflater;

    @Override
    protected void onCreate(Bundle savedInstanceState){
        super.onCreate(savedInstanceState);
        requestWindowFeature(Window.FEATURE_NO_TITLE);
        setContentView(R.layout.activity_main);
        mInflater=LayoutInflater.from(this);
        initData();
        initView();
    }

    private void initData(){
        mImgIds=new int[]{R.drawable.a, R.drawable.b, R.drawable.c,
            R.drawable.d, R.drawable.e, R.drawable.f, R.drawable.g,
            R.drawable.h, R.drawable.l};
    }

    private void initView(){
        mGallery=(LinearLayout)findViewById(R.id.id_gallery);
        for(int i=0; i<mImgIds.length; i++){
            View view=mInflater.inflate(R.layout.activity_index_gallery_item,
                    mGallery, false);
            ImageView img=(ImageView)view
                    .findViewById(R.id.id_index_gallery_item_image);
            img.setImageResource(mImgIds[i]);
            TextView txt= (TextView)view
                    .findViewById(R.id.id_index_gallery_item_text);
            txt.setText("some info ");
            mGallery.addView(view);
        }
```

    }
}

8.7 项目实战：CoffeeStore 首页广告轮播效果

8.7.1 项目分析

如图 8.9 所示，CoffeeStore 项目中的广告轮播实现一定时间内多张咖啡图进行轮流切换。该效果可以通过 Android 的 HorizontalScrollView 控件实现。

图 8.9 广告条中照片的轮转替换

8.7.2 项目实现

为实现上述效果，首先需要在布局文件中定义布局及属性，代码如下。

```
<LinearLayout xmlns:android="http://schemas.android.com/apk/res/android"
    xmlns:tools="http://schemas.android.com/tools" android:layout_width=
    "match_parent"
    android:layout_height="match_parent">
    <HorizontalScrollView
    android:layout_width="wrap_content"
    android:layout_height="150dp"
        android:layout_gravity="center_vertical"
    android:background="#AA444444"
        android:scrollbars="none">

        <LinearLayout android:id="@+id/id_gallery"
    android:layout_width="wrap_content"
        android:layout_height="wrap_content"
    android:layout_gravity="center_vertical"
        android:orientation="horizontal">
    </LinearLayout>
    </HorizontalScrollView>

</LinearLayout>
```

上面的布局文件代码通过 RadioButton 控件设置了下方按钮的效果，另外在 Java 文件中设置该图片资源的轮转替换效果，关键代码如下。

```java
public class MainActivity extends Activity {

    private LinearLayout mGallery;
    private int[] mImgIds;
    private LayoutInflater mInflater;

    @Override
    protected void onCreate(Bundle savedInstanceState){
        super.onCreate(savedInstanceState);
        requestWindowFeature(Window.FEATURE_NO_TITLE);
        setContentView(R.layout.activity_main);
        mInflater=LayoutInflater.from(this);
        initData();
        initView();

    }

    private void initData(){
        mImgIds=new int[]{R.drawable.a, R.drawable.b, R.drawable.c,
                R.drawable.d, R.drawable.e, R.drawable.f, R.drawable.g,
                R.drawable.h, R.drawable.l};
    }

    private void initView(){
        mGallery=(LinearLayout)findViewById(R.id.id_gallery);

        for(int i=0; i<mImgIds.length; i++){

            View view=mInflater.inflate(R.layout.activity_index_gallery_item,
                    mGallery, false);
            ImageView img= (ImageView)view
                    .findViewById(R.id.id_index_gallery_item_image);
            img.setImageResource(mImgIds[i]);
            TextView txt= (TextView)view
                    .findViewById(R.id.id_index_gallery_item_text);
            txt.setText("some info ");
            mGallery.addView(view);
        }
    }
```

8.7.3 项目说明

在实际项目中,通常遇到首页显示广告信息的情况,本例的广告图片切换通过 HorizontalScrollView 控件实现。除此之外,还可使用第三方插件实现这个功能,使用 org.

taptwo. android. widget. ViewFlow 和 org. taptwo. android. widget. CircleFlowIndicator 这两个插件实现此广告页。

8.8 项目实战：CoffeeStore 店铺列表页

8.8.1 项目分析

在 CoffeeStore 项目中，单击首页的店铺控件时，会进入图 8.10 所示的店铺列表页面。该页面包含了咖啡商城里全部店铺的基本信息，如店铺的名称、地址、联系方式及推荐商品等，相邻的店铺还使用了不同的显示样式。

图 8.10 店铺列表页

8.8.2 项目实现

实现这个页面，需要用到本章讲解的 ListView 控件，下面完成该店铺列表页的开发。首先在布局文件里定义该页面的布局，需要在 app/res/layout/文件夹下新建对应的布局文件 shop_layout.xml。该页面使用的是线性布局的 ListView 控件，布局文件代码如下。

```
<?xml version="1.0" encoding="utf-8"?>
<LinearLayout xmlns:android="http://schemas.android.com/apk/res/android"
    android:layout_width="match_parent"
    android:layout_height="match_parent"
    android:background="#FFFFB9"
    android:orientation="vertical">
```

```xml
<ListView
    android:id="@+id/listshop"
    android:layout_width="match_parent"
    android:layout_height="wrap_content"></ListView>
```

</LinearLayout>

定义好布局文件后,需要在对应的 Activity 文件里引用并操作该布局。其对应文件路径为 app/java/neusoft.soft.coffeestore/view/shopActivity,在该文件的 onCreate 方法中使用 setContentView 引用布局文件,具体的代码如下。

```java
public class ShopActivity extends Activity {
    List<HashMap<String, String>>data;
    final String DB_DIR="databases";
    final String DB_NAME="coffeeshop";
    private ListView list;
    ApplicationInfo applicationInfo;
    String databasePath;
    DBUtil dbUtil;
    Shop[] shops;

    @Override
    protected void onCreate(Bundle savedInstanceState){
        super.onCreate(savedInstanceState);
        setContentView(R.layout.shop_layout);
        list=(ListView)findViewById(R.id.listshop);
        init(this);
        registerForContextMenu(list);
    }

    @Override
    protected void onRestart(){
        //TODO Auto-generated method stub
        super.onRestart();
        showAllShops();
    }

    public void showAllShops(){
        shops=dbUtil.queryAllShop();
        data=new ArrayList<HashMap<String, String>>();
        for(int i=0; i<shops.length; i++){
            HashMap<String, String>map=new HashMap<String, String>();
            map.put("shop_name", shops[i].getShop_name());
            map.put("shop_address", shops[i].getShop_address());
            map.put("shop_tel", shops[i].getTel());
```

```java
            String picName=shops[i].getImg_name();
            int  picId = getResources ( ). getIdentifier (picName, " drawable ",
            ShopActivity.this.getPackageName());
            map.put("img_id", picId+"");
            data.add(map);

        }
        MyAdapter adapter=new MyAdapter
            (ShopActivity.this, data, R.layout.list_item_custom,
                new String[]{"shop_name", "shop_address", "shop_tel", "img_id"},
                new int[]{R.id.txtName, R.id.txtAddress, R.id.txtTel, R.id.img});
        list.setAdapter(adapter);
        list.setOnItemClickListener(new OnItemClickListener(){

            @Override
            public void onItemClick(AdapterView<?>arg0, View arg1, int position,
                    long arg3){
                Intent intent=new Intent();
                intent.setClass(ShopActivity.this, ShopDetailActivity.class);
                Bundle bundle=new Bundle();
                bundle.putSerializable("shop", shops[position]);
                intent.putExtras(bundle);
                startActivity(intent);
            }
        });
    }

class MyAdapter extends SimpleAdapter {

    public MyAdapter(Context context, List<?extends Map<String, ?>>data,
            int resource, String[] from, int[] to){
        super(context, data, resource, from, to);
    }
    @Override
    public View getView(int position, View convertView,
            ViewGroup parent){
        View result=super.getView(position, convertView, parent);
        TextView txtTilte=(TextView)result.findViewById(R.id.txtName);
        if(position %2 ==1){
            result.setBackgroundColor(Color.GREEN);
            txtTilte.setTextColor(Color.BLUE);
        } else {
            result.setBackgroundColor(Color.YELLOW);
            txtTilte.setTextColor(Color.RED);
```

```
        }
        return result;
    }
}
private void init(Context context){
    String packageName=context.getPackageName();
    try {
        applicationInfo= context. getPackageManager ( ). getApplicationInfo
        (packageName, PackageManager.GET_META_DATA);
        String dbDir=applicationInfo.dataDir+File.separator+DB_DIR;
        File file=new File(dbDir);
        if(!file.exists()){
            file.mkdir();
        }
        databasePath=applicationInfo. dataDir+ File. separator+ DB_DIR+ File.
        separator+DB_NAME;
    } catch(NameNotFoundException e){
    }
}
@Override
public void onCreateContextMenu(ContextMenu menu, View v,
                    ContextMenuInfo menuInfo){

    MenuInflater mInflater=getMenuInflater();
    mInflater.inflate(R.menu.shop, menu);

    super.onCreateContextMenu(menu, v, menuInfo);
}
    return super.onContextItemSelected(item);
}
```

8.8.3 项目说明

店铺列表使用 ListView 控件实现。本章的店铺列表数据定义在内存的集合里,实际项目中这些信息存储在远程数据库里。它们来源于远程数据库,通过网络通信传到本地(第 12 章中会修改 ListView 中的数据源为远程数据库)。

8.9 项目实战:CoffeeStore 首页推荐商品

8.9.1 项目分析

在 CoffeeStore 项目中,进入程序首页会出现图 8.11 所示的咖啡商城推荐商品视图,该视图将应用需要推送的商品以图文的形式显示在首页底部。推荐商品以九宫格形式显示所要推荐的商品,使用 GridView 控件实现。下面具体分析如何实现推荐商品的功能。

图 8.11　CoffeeStore 推荐商品列表

8.9.2　项目实现

　　首先在首页推荐商品区域添加一个 GridView 控件,然后定义 GridView 控件每一项使用的布局文件。分析该布局方式:使用水平排列的线性布局。第一列图片部分使用了 ImageView 控件;第二列是两行一列的布局方式,所以需要在 ImageView 控件下嵌套一个垂直排列的线性布局,具体代码如下。

```xml
<?xml version="1.0" encoding="utf-8"?>
<LinearLayout xmlns:android="http://schemas.android.com/apk/res/android"
    android:layout_width="match_parent"
    android:layout_height="match_parent"
    android:orientation="horizontal">

    <ImageView
        android:id="@+id/img"
        android:layout_width="80dp"
        android:layout_height="60dp"
        android:src="@drawable/a3" />

    <LinearLayout
        android:layout_width="wrap_content"
        android:layout_height="wrap_content"
        android:layout_marginTop="5dp"
        android:orientation="vertical">

        <TextView
            android:id="@+id/txtName"
            android:layout_width="wrap_content"
            android:layout_height="wrap_content"
            android:singleLine="true"
            android:text="TextView" />

        <TextView
            android:id="@+id/txtPrice"
            android:layout_width="wrap_content"
            android:layout_height="wrap_content"
```

```xml
            android:singleLine="true"
            android:text="TextView" />
    </LinearLayout>

</LinearLayout>
```

定义好该布局文件后,相应地在其对应的 Java 文件中引用,具体代码如下。

```java
public class HomeFragment extends Fragment {

    @SuppressLint("InflateParams")
    public View onCreateView(LayoutInflater inflater, ViewGroup container,
                        Bundle savedInstanceState){
        View view=inflater.inflate(R.layout.fragment_home, null);
        grid= (GridView)view.findViewById(R.id.grid);
        viewFlow= (ViewFlow)view.findViewById(R.id.viewflow);
        viewFlow.setAdapter(new ImageAdapter(HomeFragment.this.getActivity()));
    }
    public void showCommandCoffee(){
        final Coffee[] coffees=dbUtil.queryAllCoffee();
        List<HashMap<String, String>>data=new ArrayList<HashMap<String,
        String>>();
        for(int i=0; i<coffees.length; i++){
            HashMap<String, String>map=new HashMap<String, String>();
            map.put("coffee_name", coffees[i].getCoffee_name());
            map.put("coffee_price", coffees[i].getCoffee_price()+"");
            map.put("coffee_intro", coffees[i].getCoffee_intro());
            map.put("coffee_com", coffees[i].getCoffee_com());
            String picName=coffees[i].getImage_name();
            int picId = getResources ( ). getIdentifier (picName, " drawable ",
            HomeFragment.this.getActivity().getPackageName());
            map.put("img_id", picId+"");
            data.add(map);
        }
        SimpleAdapter adapter=new SimpleAdapter
                (HomeFragment.this.getActivity(), data, R.layout.gridview_item_layout,
                    new String[]{"coffee_name", "coffee_price", "img_id"},
                    new int[]{R.id.txtName, R.id.txtPrice, R.id.img});
        grid.setAdapter(adapter);
    }
}
```

8.9.3 项目说明

首页的推荐商品采用九宫格的形式显示,这就用到了 Android 的 GridView 控件。它与 Listview 的用法相似,只是数据的显示形式与 ListView 不同,不是以列表的形式显示,而是以九宫格的形式显示。

本章小结

本章主要介绍了 Android 常用高级控件的用法,包括 Spinner 控件、ListView 控件、GridView 控件、HorizontalScrollView 等控件。通过店铺列表页、图片的切换以及首页分类条等功能的实现,全面介绍了各个控件的使用方法及可能出现的问题。

Android 常用高级控件在程序开发中占有十分重要的地位。与前面章节相比,高级控件的应用会更加复杂,比如 ListView 控件和 ExpandableListView 控件。特别是涉及 Adapter 控件及其相关使用方法时,难度明显增大。读者应多加练习,以便熟练掌握并使用这些高级控件,为下一步学习奠定基础。

本章习题

1. 下面关于 Adapter 的描述,错误的一项是(　　)。
   A. Android 系统提供了几个默认的 Adapter 类供开发者使用,开发者也可以继承 Adapter 类来自定义 Adapter
   B. Adapter 对象在 Adapter 控件和数据源之间扮演桥梁的角色,它提供了访问数据源的入口,并把从数据源拿到的数据逐项加载到 Adapter 控件中
   C. Android 有 4 种 Adapter 对象可供开发者使用,分别是 ArrayAdapter、SimpleAdapter、SimpleCursorAdapter 和自定义的 Adapter
   D. Android 使用了一个抽象类——BaseAdapter 作为各个 Adapter 实体类的基类,并使用两个接口——ListAdapter 和 SpinnerAdapter 分别作为两种类型的 AdapterView——AbsListView(包含 ListView 和 GridView)和 AbsSpinner(包含 Spinner 和 Gallery)的适配接口

2. 下列关于 Adapter 的继承关系,描述错误的一项是(　　)。
   A. SimpleAdapter 继承自 BaseAdapter
   B. CursorAdapter 继承自 BaseAdapter
   C. ArrayAdapter 继承自 BaseAdapter
   D. 自定义 Adapter 继承自 Adapter

3. 下列关于 ListView 的特点,描述错误的是(　　)。
   A. 采用 MVC 模式将前端显示和后端数据分离
   B. 为 ListView 提供数据的 List 或数组相当于 MVC 模式中的 M(数据模型 Model)
   C. ListView 相当于 MVC 模式中的 V(视图 View)
   D. ListAdapter 对象相当于 MVC 模式中的 C(控制器 Control)

4. 下列关于 ListView 的 XML 属性,描述错误的一项是(　　)。
   A. ListView 与其他的 UI 控件相同,在 XML 布局文件中通过<ListView>标签

将其放入界面布局中

B. 直接让 Activity 继承自 ListAdapter,可以将 ListView 填充满整个 Activity

C. 在 XML 布局代码中将 ListView 的位置设为占满整个 Activity,可以将 ListView 填充满整个 Activity

D. 可以把 ListView 放在布局控件中,让其只占界面的某一部分

5. 下列关于 4 种类型的 Adapter 的描述,错误的一项是(　　)。

A. ArrayAdapter 是最简单的 Adapter,适合于列表项中只含有文本信息的情况,是填充文本列表最简便的一种方式

B. SimpleAdapter 比 ArrayAdapter 复杂,适合于每一个列表项中含有不同的子控件,只能是含有一个图片和一串文本的组合

C. SimpleCursorAdapter 专门用来把一个 Cursor(游标)中的数据映射到列表中,Cursor 中的每一条数据映射为列表中的一项

D. 自定义 Adapter 是完全自行定义数据的适配方式,灵活性最强,但使用起来比前 4 个复杂

6. 下列关于 ListView 的描述,错误的一项是(　　)。

A. ListView 是一个列表,用户自然需要选择其中的某一项做一些处理。为了响应用户的单击事件,可以使用 setOnItemClickListener()方法

B. ListView 的单击事件处理,一般都使用一个匿名内部类对象来处理

C. 在 onItemLongClick()方法中,id 表示被长按的列表项在 ListView 中的位置

D. ListView 长按某一个列表项,需要使用 setOnItemLongClickListener()方法

7. 下列说法,错误的一项是(　　)。

A. 可以使用 ArrayAdapter 向一个 ListView 中填充数据

B. 可以使用 SimpleAdapter 向一个 ListView 中填充数据

C. 不可以使用 SimpleCursorAdapter 向一个 ListView 中填充数据

D. 可以使用自定义 Adapter 向一个 ListView 中填充数据

8. 下列对于常用的 AdapterView 的描述,错误的一项是(　　)。

A. ListView 和 GridView 同属于 AbsListView 的子类,两者具有相似性

B. Spinner 也表现为一种列表,它的主要作用是让用户进行选择,比如一个下拉列表

C. GridView 是一个表格显示资源的控件,其上显示的资源只能使用 ArrayAdapter 填充

D. Gallery 是一个显示缩略图的控件,它可以把子项显示在一个中心锁定且水平滚动的列表中

9. 请写出 Android 中可供开发者使用的 4 种 Adapter 对象。

10. 请至少写出 ListView 的 3 个特点。

11. 请写出常用的 4 种类型的 Adapter 及其各自的特点。

12. CoffeeStore 商品列表页如何实现?如何从店铺列表跳到商品列表?请简述其方法。

13. 在 CoffeeStore 项目中,如果从店铺列表跳到商品列表页时需要将用户单击的咖啡商品信息显示到页面上,那么在 Activity 跳转过程中如何进行传值操作?请简述之。

资源样式与主题

本章概述

一个Android应用需要使用各种资源文件,这些文件可以是一些具体属性(文本、颜色、尺寸),多媒体文件(图片、音频、视频、动画),也可以是美化程序的样式资源和主题资源。这里所说的并不是程序本身的资源,而是通过程序代码引用的外部资源或自定义的资源文件。一般来说,资源可以看做是可以复用的、只读的数据资源。本章重点讲述Android开发中的资源文件。

学习重点与难点

重点:

(1) 值资源。
(2) 位图资源与色图资源。
(3) XML资源。
(4) 菜单资源。
(5) 对话框资源。
(6) 动画资源。
(7) 风格资源与主题。

难点:

(1) 位图资源与色图资源。
(2) 动画资源及单击动画。
(3) 选择器。

学习建议

本章涉及的资源文件是容易忽略的内容,读者往往感觉学习难度不大,但应用中往往会因为不注意而产生各种各样的错误。本章应重点掌握不同资源文件存放的位置、资源的属性、调用的方法等内容。熟练掌握并使用字符串资源、图片资源、值资源、风格与主题资源、动画等资源,是Android开发的重中之重。

9.1 资　　源

在Android Studio中,资源文件会保存到res目录下,主要的资源类型有值资源(dimens/strings/styles/array/color)、布局资源(layout)、图片资源(drawable)与菜单资

源(menu)。具体的保存位置如表 9.1 所示。

表 9.1 Android 中的常用资源

位　　置	放置内容
app/res/values	字符串、颜色、尺寸、数组、样式、主题资源等
app/res/layout	布局资源
app/res/drawable	图片资源，也可以使用多种 XML 文件作为资源
app/res/anim	动画资源
app/res/menu	菜单资源

上述文件夹中的资源并不是程序本身的资源，而是通过程序代码引用的外部资源或自定义的资源文件，所以新建工程时并不会自动生成所有的对应文件，需要手动添加。添加时找到放置资源文件的文件夹，右击，选择 New 添加，同时开发环境会在 R.java 文件中自动生成对应的资源名称，需要使用对应的资源文件时，只需在开发环境中调用即可。这些资源可以在 Java 文件中使用，也可以在其他 XML 资源中使用，一般推荐在 XML 资源中使用。该资源编译时会被编译到应用程序中去。使用程序时，资源文件会作为程序的一部分进行调用。

资源文件的使用包括以下两种方式。

1. 在资源中引用资源

@[package:]type:name

- @表示对资源的引用，@标识表示需要解析的内容而非要显示的内容。
- package 是包名称，如果在相同的包，package 则可以省略。

2. 在代码中引用资源

在代码中引用资源，需要使用资源的 ID，可以通过[R.resource_type.resource_name]或[android.R.resource_type.resource_name]获取资源 ID。

- resource_type：代表资源类型，即 R 类中的内部类名称。
- resource_name：代表资源名称，对应资源的文件名或在 XML 文件中定义的资源名称属性。

下面通过例 9-1 实现在资源中引用资源。

【**例 9-1**】 资源文件使用。

首先对 app/res/values 下的 strings.xml 文件进行修改，添加一条名称为 my_string_resource_a 的字符串资源，使用第一种方式引用该资源；同时定义 my_string_resource_b 的字符串资源，使用第二种方式引用该资源。

```
<resources>
    <string name="app_name">Example-09-1</string>
    <string name="my_string_resource_a">My String Resource A</string>
```

```xml
<string name="my_string_resource_b">My String Resource B</string>
</resources>
```

之后在主布局文件中建立两个 TextView 控件,并设置其属性,将第一个 TextView 的 android:text 属性设置为引用资源文件,第二个 TextView 控件不设置 android:text 属性。代码如下。

```xml
<TextView
    android:layout_width="wrap_content"
    android:layout_height="wrap_content"
    android:layout_margin="2px"
    android:text="@string/my_string_resource_a" />

<TextView
    android:id="@+id/tvB"
    android:layout_width="wrap_content"
    android:layout_height="wrap_content"
    android:layout_margin="2px" />
```

此时,在 MainActivity 中绑定布局文件,就可以显示字符串资源 my_string_resource_a 的值。如果在 Activity 中显示 my_string_resource_b 的值,还需要编写代码,首先调用 findViewById 方法,找到 ID 为 tvB 的 TextView 控件,然后调用 this.getResources().getString()方法,找到对应的字符串资源,最后调用 setText 方法实现字符串资源 my_string_resource_b 的显示。具体代码如下。

```java
private TextView tvB;

@Override
protected void onCreate(Bundle savedInstanceState){
    super.onCreate(savedInstanceState);
    setContentView(R.layout.activity_main);
    tvB=(TextView)findViewById(R.id.tvB);
    String resourceValue= this.getResources().getString(R.string.my_string_resource_b);
    tvB.setText(resourceValue);
```

9.2 值 资 源

9.2.1 字符串资源

字符串是使用频繁的一类资源,放置在 app/res/values/strings.xml 文件中。将字符串声明在配置文件中,更加有利于布局和修改,另外对程序的国际化有很大帮助。该文件中可以存放的字符串类型如下。

- 普通的字符串。

- 引用字符串。
- HTML 字符串(引用字符串与 HTML 字符串应更改 XML 文件的内容)。
- 可替换字符串(只能使用代码方式获取资源值后再赋值)。

例 9-1 中使用的是普通字符串资源,例 9-2 中使用的是其他类型的字符串资源。

【例 9-2】 定义不同类型字符串资源并显示。

首先在 strings.xml 中使用如下代码。

```
<resources>
    <string name="app_name">Example-09-2</string>
    <string name="common_string">普通字符串</string>
    <string name="quoted_string">\"引用字符串\"</string>
    <string name="html_string"><b><i>HTML</i></b><b>字符串</b></string>
</resources>
```

在布局文件中使用 TextView 控件,分别引用不同的字符串资源,具体代码如下。

```
<TextView
android:layout_width="fill_parent"
android:layout_height="wrap_content"
android:layout_margin="2px"
android:text="@string/common_string" />
<TextView
android:layout_width="fill_parent"
android:layout_height="wrap_content"
android:layout_margin="2px"
android:text="@string/quoted_string" />
<TextView
android:layout_width="fill_parent"
android:layout_height="wrap_content"
android:layout_margin="2px"
android:text="@string/html_string" />
<TextView
android:layout_width="fill_parent"
android:layout_height="wrap_content"
android:layout_margin="2px"
android:text="@string/java_format_string" />
```

其中前三个字符串资源比较容易实现。对于可替换字符串来说,只能使用代码方式获取资源值后再赋值。所以需要在 Java 文件中进行设置。首先在布局文件中设置可替换字符串 id 属性为 txtJavaFormatSting,在 Java 文件中使用如下代码。

```
txtJavaFormatSting= (TextView)(this.findViewById(R.id.txtJavaFormatSting));
String value = String.format (this.getResources().getString (R.string.java_format_string),"TOM");
txtJavaFormatSting.setText(value);
```

最终运行效果如图 9.1 所示。

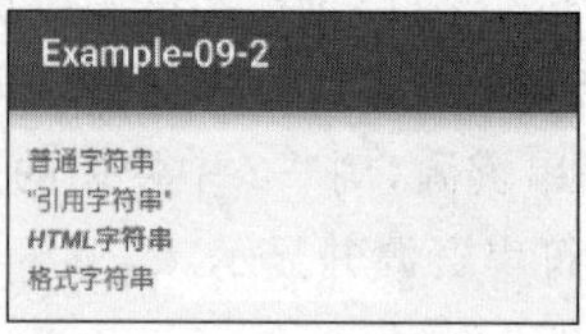

图 9.1　字符串资源使用

9.2.2　颜色资源

对 Android 程序来说，颜色资源是最基本、最常用的资源，它可以对文本、背景、组件的部件进行设置。颜色资源位于 app/res/values/colors.xml 文件中，该文件可以存放的颜色资源类型如下。

- RGB 值，是通过对红（R）、绿（G）、蓝（B）三个颜色通道的变化以及它们相互之间的叠加来得到各式各样的颜色的，是目前运用最广的颜色系统之一。
- ARGB 值（如果 ARGB 每个值的两位 16 进制值相同，可以省略为 1 个。

比如红色为＃FFFF0000，可写成＃FF00 或＃F00）。常用的颜色资源有＃FFFF0000 红色、＃FFFF6600 橙色、＃FFFFEE00 黄色、＃0f0 绿色，具体颜色对照表见附录 A。

在颜色资源文件中定义颜色资源后，可以在 Java 或 XML 文件中使用该颜色资源。颜色资源 XML 文件中的使用语法格式如下。

@[<package>:]color/颜色资源名

下面通过一个例子演示引用颜色资源及其使用方法。

【例 9-3】　对例 9-2 的扩展：为 TextView 控件添加颜色资源。

首先在 app/res/values/colors.xml 文件中定义颜色资源，代码如下。

```
<?xml version="1.0" encoding="utf-8"?>
<resources>
    <color name="colorPrimary">#3F51B5</color>
    <color name="colorPrimaryDark">#303F9F</color>
    <color name="colorAccent">#FF4081</color>
    <color name="red">#FF0000</color>
    <color name="green">#00FF00</color>
</resources>
```

定义完成颜色资源文件后，在布局文件中定义 TextView 组件时，为 TextView 组件添加 Android：textColor 属性，通过引用的方式引用资源文件，关键代码如下。

```
<TextView
    android:layout_width="wrap_content"
    android:layout_height="wrap_content"
```

```
android:layout_margin="2px"
android:textColor="@color/red"
android:text="@string/common_string" />
```

这样就可以通过引用资源文件将 TextView 中的字符串显示颜色设置成红色。

9.2.3 尺寸资源

尺寸资源也称为 dimen 资源,可以设置文本,组件的大小、宽度、高度、间距等。Android 中的显示单位如下。

- px(pixels)像素,此像素用得比较多,1px 代表屏幕上一个物理像素点,一般 HVGA 代表 320×480 像素。
- dip 或 dp(device independent pixels),设备独立像素。
- 该单位和设备硬件有关,为了支持 WVGA、HVGA 和 QVGA,一般推荐使用该单位,不依赖像素。
- sp(scaled pixels-best for text size)比例像素,主要处理字号的大小,可以根据系统的字号自适应。

除了上面的尺寸资源以外,下面列举了一些不太常用的尺寸文件。

- in(inches)英寸。
- mm(millimeters)毫米。
- pt(points)点,1/72 英寸。

在人机交互设计课程中,使用的长度单位是 px。px 代表像素点,对一般网页,显示的区别不是特别大,但是对于不同的手机设备,不同机型的像素差别较大,会导致同样像素的资源在不同设备上显示的大小完全不同。为了适应不同的分辨率和像素密度,对于控件的尺寸,推荐使用 dip ,对于文字的尺寸,推荐使用 sp。

【例 9-4】 下面使用一个实例说明如何使用尺寸资源。

首先在 dimen.xml 文件中设置尺寸信息。

```
<resources>
    <dimen name="activity_horizontal_margin">16dp</dimen>
    <dimen name="activity_vertical_margin">16dp</dimen>
    <dimen name="common_text_size">30sp</dimen>
</resources>
```

同时,对布局文件的 TextView 控件使用该尺寸资源,代码如下。

```
<TextView
    android:layout_width="wrap_content"
    android:layout_height="wrap_content"
    android:text="Dimen 测试 1" />
<TextView
    android:layout_width="wrap_content"
    android:layout_height="wrap_content"
```

```
android:textSize="@dimen/common_text_size"
android:text="Dimen测试2" />
```

最终程序运行效果如图 9.2 所示。

图 9.2　尺寸资源使用

9.2.4　数组资源

数组资源位于 app/res/values 目录下，通过数组资源可以存放一维的字符串数组或整数数组。定义时使用 string-array 元素定义字符串数组，使用 integer-array 定义整数数组。定义数组资源时使用 name 属性，并需要在起始标记和结束标记中使用＜item＞＜/item＞标签对定义单个数组元素。

【例 9-5】　定义一个数组，具体显示 Beijing、Shanghai、Tianjin、Dalian 等信息，使用如下代码。

```
<string-array name="listitem">
<item>Beijing</item>
<item>Shanghai</item>
<item>Tianjin</item>
<item>Dalian</item>
</string-array>
```

下面使用简单的 ListView 控件显示上述数组，在主布局文件中使用字符串数组资源，为其指定 android：entries 属性。关键代码如下。

```
<ListView
    android:layout_width="wrap_content"
    android:layout_height="wrap_content"
    android:entries="@array/listitem" />
```

最终显示效果如图 9.3 所示。

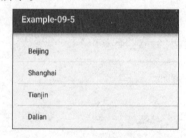

图 9.3　数组资源使用

9.3 位图资源与色图资源

位图资源即图片,存放在 app/res/drawable 目录中。其中根据屏幕的分辨率设置 3 个级别的资源：drawable-hdpi 保存高分辨率的图片,drawable-mdpi 保存中等分辨率的图片,而 drawable-ldpi 保存低分辨率的图片。

色图资源即某种纯色的"图片"资源,存放在 app/res/values 目录中,同样也可以设置不同分辨率下的资源。

使用图片资源时,主要包括以下几种格式的图片资源,如表 9.2 所示。

表 9.2　Android 中图片资源使用情况

名　　称	扩展名	描　　述
便携式网络图像(portable network graphics)	png	无损的图片格式,推荐使用
9 格拉伸图片(nine patch strechable images)	9.png	由 png 格式转换而来,无损,推荐使用
联合图像专家组(joint photographics experts group)	jpg	有损的图片格式,可以接受,不推荐使用
图像交换形式(graphics interchange format)	gif	使用频率较低

这里重点介绍 9Patch(9 格拉伸)图片。9Patch 是 png 格式图片的一种变种,它最大的特点就是可拉伸性,适用于 Android 手机应用程序的背景图,使用时与常规的 drawable 资源相同。图中的黑点表示可拉伸的位置,下面的黑色线表示文本内容的范围。

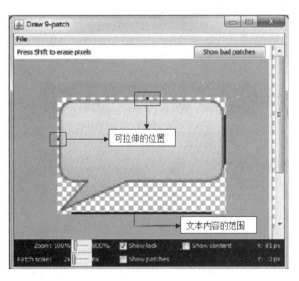

图 9.4　9Patch 图片资源

【例 9-6】 9 格拉伸图片资源。

本例介绍 9 格拉伸图片的使用。首先将图片资源粘贴到程序的 drawable 文件夹下,图片命名为 bak.9.png,在布局文件中使用 2 个 TextView 控件,对第 1 个控件直接引用图片资源,对第 2 个控件使用 style 属性引用图片资源,本例布局文件的代码如下。

```
<TextView
    android:layout_width="wrap_content"
    android:layout_height="wrap_content"
    android:layout_margin="10sp"
    android:background="@drawable/bak"
    android:text="Hello!"
    android:textSize="40sp" />

<TextView
    style="@style/chart_style"
    android:text="Hello,Nice to see you!"
    android:textSize="40sp" />
```

以上代码实现了对第 1 个控件直接引用图片资源。而对第 2 个控件使用 style 属性引用图片资源,style 属性的代码如下。

```
<resources xmlns:android="http://schemas.android.com/apk/res/android">

    <style name="chart_style">
        <item name="android:layout_width">wrap_content</item>
        <item name="android:layout_height">wrap_content</item>
        <item name="android:background">@drawable/bak</item>
        <item name="android:textColor">#F00</item>
    </style>
</resources>
```

最终显示效果如图 9.5 所示。

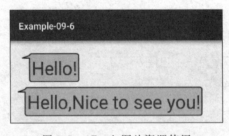

图 9.5　9Patch 图片资源使用

9.4　XML 资源

本节所述的 XML 资源与平时所说的 XML 文件不同,该 XML 资源是以 XML 形式保存的数据资源,不是定义资源文件时使用的 XML 文件。该 XML 文件数据存放在 app/res/xml 文件夹中。在 Android 项目中,一般需要手动创建该 XML 文件,并且使用

XmlPullParser 解析数据。

【例 9-7】 建立和使用 XML 资源。

使用 XML 资源文件。首先修改布局文件,在该布局文件中使用 TextView 组件,设置组件 id 为 show。其次,在 res 目录中创建新的 XML 文件,在该文件中添加根节点及子节点,这些节点用来保存信息,具体的 XML 文件代码如下。

```
<?xml version="1.0" encoding="utf-8"?>
<customers>
<customer
name="BeiJing"
tel="010" />
<customer
name="Tianjin"
tel="022" />
<customer
name="Shanghai"
tel="021" />
<customer
name="Dalian"
tel="0411" />
</customers>
```

下面需要在 MainActivity 中获取 XML 文档,并使用 while 循环遍历该文档,最后将获取到的结果显示在该控件上,运行代码如下。

```
XmlResourceParser xrp=getResources().getXml(R.xml.customer);
StringBuilder sb=new StringBuilder("");
try {
while(xrp.getEventType()!=XmlResourceParser.END_DOCUMENT){
if(xrp.getEventType()==XmlResourceParser.START_TAG){
String tagName=xrp.getName();
if(tagName.equals("customer")){
sb.append("城市"+xrp.getAttributeValue(0)+"   ");
sb.appcnd("区号"+xrp.getAttributeValue(1)+"   ");
sb.append("\n");
}
    }
    xrp.next();
}
    TextView tv=(TextView)findViewById(R.id.show);
tv.setText(sb.toString());
} catch(XmlPullParserException e){
    e.printStackTrace();
} catch(IOException e){
```

```
        e.printStackTrace();
    }
```

最终显示效果如图9.6所示。

```
城市BeiJing 区号010
城市Tianjin 区号022
城市Shanghai 区号021
城市Dalian 区号0411
```

图9.6 使用XML文件显示具体内容

9.5 菜单资源

在Android程序中,菜单资源应用广泛,它位于app/res/menu目录下。菜单主要分为选项菜单(ActionBar)和上下文菜单(Content menu)。选项菜单就是单击手机中Menu弹出的菜单,而上下文菜单类似于Windows下鼠标右击弹出的快捷菜单,在手机中长按已经设置好的控件,可以弹出对应的上下文菜单。

1. 选项菜单

【例9-8】 定义和使用菜单。

菜单可以使用代码来定义,也可以使用XML文件来定义,下面分别介绍两种菜单定义方法。

1) 使用XML菜单文件

菜单资源位于app/res/menu目录下,新建工程时一般没有这个文件夹,直接在res文件夹下新建一个menu文件夹即可,该文件夹下存放程序需要的菜单资源。创建时在menu文件夹下右击,在弹出的快捷菜单中选择new下的Menu Resourse file即可。接下来在XML文件中设置,XML菜单的代码如下。

```xml
<?xml version="1.0" encoding="utf-8"?>
<menu xmlns:android="http://schemas.android.com/apk/res/android"
    xmlns:app="http://schemas.android.com/apk/res-auto">
    <group android:id="@+id/menu_group">
        <item android:id="@+id/index"
            app:showAsAction="always"
            android:icon="@drawable/index"
            android:title="主页"/>
        <item android:id="@+id/weather"
            app:showAsAction="ifRoom"
            android:icon="@drawable/weather"
            android:title="天气">
            <menu>
                <item android:id="@+id/today"
```

```xml
                android:title="今日天气"/>
            <item android:id="@+id/week"
                android:title="一周天气"/>
        </menu>
    </item>
    <item android:id="@+id/trend"
        app:showAsAction="ifRoom"
        android:icon="@drawable/trend"
        android:title="趋势"/>
    <item android:id="@+id/tools"
        app:showAsAction="ifRoom"
        android:icon="@drawable/tools"
        android:title="工具"/>
</group>
</menu>
```

这里定义了一个名为 menu_main 的菜单资源文件，代码中每个 item 代表一个选项。同时会存在 menu 里嵌套 menu 的情况。在 Android 开发中，不光可以设置菜单资源，也可以为菜单添加子菜单，子菜单也使用<menu>标签声明，内部使用<item>标签描述菜单项。接下来在 Activity 中将定义好的菜单资源引入程序，首先重写 onCreateOptionsMenu()方法，在该方法中引入定义好的菜单资源，使用 getMenuInflater().inflate(R.menu.menu_main,menu);语句，这里 inflate()方法引入了刚刚定义的 menu_main 资源，具体代码如下。

```java
public class MenuActivity extends AppCompatActivity {
    @Override
    protected void onCreate(Bundle savedInstanceState){
        super.onCreate(savedInstanceState);
        setContentView(R.layout.activity_menu);
    }
    public boolean onCreateOptionsMenu(Menu menu){
        getMenuInflater().inflate(R.menu.menu_main,menu);
        return super.onCreateOptionsMenu(menu);
    }
    @Override
    public boolean onOptionsItemSelected(MenuItem item){
        switch(item.getItemId()){
            case R.id.index:
                Toast.makeText(this,"您单击的是主页",Toast.LENGTH_SHORT).show();
                break;
            case R.id.weather:
                Toast.makeText(this,"您单击的是 weather",Toast.LENGTH_SHORT).show();
                break;
            case R.id.trend:
```

```
                Toast.makeText(this,"您单击的是 trend",Toast.LENGTH_SHORT).show();
                break;
            case R.id.tools:
                Toast.makeText(this,"您单击的是 tools",Toast.LENGTH_SHORT).show();
                break;
        }
        return super.onOptionsItemSelected(item);
    }
    @Override
    public boolean onMenuOpened(int featureId, Menu menu){
        if(menu!=null){
            if(menu.getClass().getSimpleName().equalsIgnoreCase("MenuBuilder")){
                try {
                    Method method=menu.getClass().getDeclaredMethod
                        ("setOptionalIconsVisible", Boolean.TYPE);
                    method.setAccessible(true);
                    method.invoke(menu, true);
                } catch(Exception e){
                    e.printStackTrace();
                }
            }
        }
        return super.onMenuOpened(featureId, menu);
    }
}
```

上述代码使用了选项菜单。单击菜单项需要使用 onOptionsItemSelected()方法设置单击事件,为了完全显示菜单内容,还需要定义 onMenuOpened()方法,否则菜单中的图片无法正常显示。最终显示效果如图9.7所示。

2) 使用代码定义菜单

除了定义 menu 文件夹下的菜单资源文件外,也可以使用代码定义菜单。使用 menu 对象的 add 方法添加菜单项,重新定义 onCreateOptionsMenu()方法,这种方法也可以实现菜单效果,具体代码如下。

图9.7 显示选项菜单

```
public boolean onCreateOptionsMenu(Menu menu){
    MenuItem index=menu.add(Menu.NONE, Menu.FIRST, Menu.FIRST, "主页");

    index.setIcon(R.drawable.index).setShowAsAction(MenuItem.SHOW_AS_ACTION_IF_ROOM);
    SubMenu wea=menu.addSubMenu(Menu.NONE, Menu.FIRST+1, Menu.FIRST+1,
    "天气").
        setIcon(R.drawable.weather);
```

```
wea.add(Menu.NONE,Menu.FIRST+5,Menu.FIRST+5,"今日天气");
wea.add(Menu.NONE,Menu.FIRST+6,Menu.FIRST+6,"一周天气");
menu.add(Menu.NONE,Menu.FIRST+2,Menu.FIRST+2,"趋势").setIcon(R.drawable.
    trend);
menu.add(Menu.NONE,Menu.FIRST+3,Menu.FIRST+3,"工具").setIcon(R.drawable.
    tools);
menu.add(Menu.NONE,Menu.FIRST+4,Menu.FIRST+4,"设置").setIcon(R.drawable.
    setting);
return super.onCreateOptionsMenu(menu);
}
```

这里使用了 add()方法,其中的 4 个参数介绍如表 9.3 所示(以第一行"主页"为例)。

表 9.3　menu 对象的 add 方法参数介绍

用　　例	放　置　内　容
Menu.NONE	组 Id
Menu.FIRST	菜单项 Id
Menu.FIRST	排序 Id
主页	字符串类型或资源 Id(int)类型的菜单项内容

需要注意的是,选项菜单使用 Activity 的 onCreateOptionMenu()方法关联,应用程序最多显示 6 个选项菜单,超出 6 个时,程序会自动显示 More 按钮。

2. 上下文菜单

上下文菜单的定义与选项菜单类似,都是在 Activity 的 onCreate()中注册,并在 Activity 的 onCreateContextMenu()方法中绑定。上下文菜单需要和具体控件绑定,长按该控件时实现上下文菜单的效果。为例 9-8 添加功能,加入一个 id 为 tv 的 TextView 控件,对该控件绑定上下文菜单,菜单中定义红色、绿色、蓝色 3 个条目,单击对应条目时,将更换 TextView 的背景颜色。

首先定义一个新的菜单资源 main_menu,代码如下。

```
<?xml version="1.0" encoding="utf-8"?>
<menu xmlns:android="http://schemas.android.com/apk/res/android">
    <item android:id="@+id/red"
        android:title="红色"
        android:orderInCategory="1">
    </item>
    <item
        android:title="绿色"
        android:id="@+id/green"
        android:orderInCategory="2"
        >
```

```xml
        </item>
        <item
            android:title="蓝色"
            android:id="@+id/blue"
            android:orderInCategory="3"
            >
        </item>
    </menu>
```

在布局文件中定义一个 id 为 tv 的 TextView 控件,在 Activity 中绑定注册时,需要对该控件使用 registerForContextMenu()方法。弹出菜单后,单击每个条目需要重写 onContextItemSelected()方法,实现代码如下。

```java
public class MenuActivity extends AppCompatActivity {
    private TextView tv;
    @Override
    protected void onCreate(Bundle savedInstanceState){
        super.onCreate(savedInstanceState);
        setContentView(R.layout.activity_menu);
        tv=(TextView)findViewById(R.id.tv);
        registerForContextMenu(tv);
    }
    @Override
    public void onCreateContextMenu(ContextMenu menu, View v, ContextMenu.ContextMenuInfo menuInfo){
        getMenuInflater().inflate(R.menu.main_menu,menu);
        super.onCreateContextMenu(menu, v, menuInfo);
    }
    @Override
    public boolean onContextItemSelected(MenuItem item){
        switch(item.getItemId()){
            case R.id.red:
                tv.setBackgroundColor(Color.RED);
                break;
            case R.id.green:
                tv.setBackgroundColor(Color.GREEN);
                break;
            case R.id.blue:
                tv.setBackgroundColor(Color.BLUE);
                break;
        }
        return super.onContextItemSelected(item);
    }
}
```

单击条目后设置文字的背景颜色,最终的显示代码如图 9.8 所示。

图 9.8　显示上下文菜单

某些控件已带有上下文菜单,默认情况下,自定义菜单将与已定义的菜单合并,也可以使用 menu 对象的 clear() 方法清空原有菜单项。

9.6　对话框资源

9.6.1　提醒(Toast)对话框

Toast 是一种轻量级控件。它的名称十分形象,用 Toast(吐司面包)表示控件效果出现后又逐渐恢复的效果。单击该控件会弹出提示性信息,一段时间后又逐渐消失。不过该控件被单击后弹出提示信息,并不参与用户交互。

【例 9-9】　下面使用 Toast 显示文本效果,主要功能是设置一个按钮,单击该按钮后 Toast 弹出提示信息。其 XML 文件代码如下。

```
<?xml version="1.0" encoding="utf-8"?>
<RelativeLayout xmlns:android="http://schemas.android.com/apk/res/android"
    xmlns:tools="http://schemas.android.com/tools"
    android:layout_width="match_parent"
    android:layout_height="match_parent"
    android:paddingBottom="@dimen/activity_vertical_margin"
    android:paddingLeft="@dimen/activity_horizontal_margin"
    android:paddingRight="@dimen/activity_horizontal_margin"
    android:paddingTop="@dimen/activity_vertical_margin"
    tools:context="com.example.administrator.myapplication.MainActivity">

    <Button
        android:id="@+id/button"
        android:layout_width="wrap_content"
        android:layout_height="wrap_content"
        android:layout_alignParentEnd="true"
```

```
android:layout_alignParentStart="true"
android:layout_alignParentTop="true"
android:onClick="viewText"
android:text="显示文本" />
</RelativeLayout>
```

这里只是建立了一个按钮，按钮显示文字为"显示文本"，按钮名称为 button。下面在 Activity 文件中实现单击该按钮时显示 Toast 效果。具体代码如下。

```
public class MainActivity extends AppCompatActivity {
private View v;

@Override
protected void onCreate(Bundle savedInstanceState){
super.onCreate(savedInstanceState);
//绑定布局文件,这样这两个文件就管理起来了,通过这个类使两个文件关联。
setContentView(R.layout.activity_main);
}

//下面要自定义一个方法,按钮的单击事件方法。
public void viewText(View v){
        Toast.makeText(getApplicationContext(),"此时按钮被单击,这里是 Toast 显示的效果", Toast.LENGTH_LONG).show();
}
```

上面通过 Toast 显示了文字效果，也可以使用 Toast 显示图片，甚至是图片和文字同时显示。下面在上例 XML 文件的基础上添加代码，实现图片和图文的显示效果。XML 文件添加代码如下。

```
<Button
android:id="@+id/button2"
android:layout_alignParentEnd="true"
android:layout_alignParentStart="true"
android:layout_below="@+id/button"
android:layout_height="wrap_content"
android:layout_width="wrap_content"
android:onClick="viewImage"
android:text="显示图像" />

<Button
android:layout_width="wrap_content"
android:layout_height="wrap_content"
android:text="显示图文"
android:id="@+id/button3"
android:onClick="viewImageText"
```

```
android:layout_below="@+id/button2"
android:layout_alignParentEnd="true"
android:layout_alignParentStart="true" />
</RelativeLayout>
```

在 activity 中设置图片显示的代码如下。

```
public void viewImage(View v){                    //用于显示图片的组件
Toast t=new Toast(this);
ImageView imageView=new ImageView(this);          //很多组建立都有 context 参数
        //为图片组件设置图片
imageView.setImageResource(R.drawable.picture);
                //为符合 java 命名规范,所有的图片资源必须用小写,不能以数字打头。
t.setView(imageView);
t.setDuration(Toast.LENGTH_LONG);
t.setGravity(Gravity.TOP,0,0);    //设置显示位置,第一个参数是位置,后面 2 个是偏移量
t.show();
}
```

通过设置图文显示的代码如下。

```
public void viewImageText(View v){
        Toast t=new Toast(this);
TextView textView=new TextView(this);
textView.setText("TDHFDJSFHDKJS");
ImageView imageView=new ImageView(this);
imageView.setImageResource(R.drawable.picture);
LinearLayout layout=new LinearLayout(this);
layout.setOrientation(LinearLayout.VERTICAL);
layout.setGravity(Gravity.CENTER);
layout.addView(imageView);
layout.addView(textView);
t.setView(layout);
t.setGravity(Gravity.CENTER,0,0);
t.setDuration(Toast.LENGTH_LONG);
t.show();
```

在上面的 activity 代码中,可以通过 Toast 对象的方法定义复杂的提醒,举例如下。
- setView 方法,设置 Toast 显示对应的 View。
- setGravity 方法,设置 Toast 对应的位置,layout. setGravity(Gravity. CENTER);即居中显示。

最终显示效果如图 9.9 所示。

对话框提供一种与用户交互的功能。在 Android 中,对话框是以异步的模态显示的。所谓模态对话框,就是这个对话框弹出时,鼠标不能单击这个对话框之外的区域,该对话框往往是用户进行了某种操作后才出现的。非模态对话框通常用于显示用户经常访问的

(a) Toast 控件显示文字效果　　　　　(b) Toast 控件显示文字和图片效果

图 9.9　Toast 控件显示效果

控件和数据,并且使用这个对话框的过程中需要访问其他窗体的情况。

对话框主要有简单的提醒对话框、带多选的对话框、自定义布局对话框、进度对话框、时间对话框、日期对话框。它们之间的关系如图 9.10 所示。

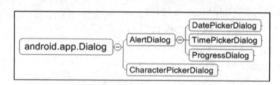

图 9.10　几种典型的对话框

下面分别介绍典型的对话框资源。

9.6.2　AlertDialog

在 Android 中,对话框需要使用 AlertDialog 类来创建,该类的主要功能是弹出一个提醒消息。但是 AlertDialog 类并没有 public 的构造方法,所以需要使用 Builder 类来构造,AlertDialog.Builder 对象构建对话框,在此基础上可以添加对话框的确认按钮及对应的事件处理。

【例 9-10】提醒对话框示例。

下面构造一个最简单的提醒对话框。实现方法很简单,需要在 Activity 中实现,代码如下。

```
private void btnSimpleAlertDialog_OnClick(View v){
    AlertDialog.Builder builder=new AlertDialog.Builder(this);
builder.setTitle("系统信息");
```

```
builder.setIcon(R.drawable.faq);
builder.setMessage("您看到的是提醒对话框");

DialogInterface.OnClickListener()=
new DialogInterface.OnClickListener(){
@Override
publicvoid onClick(DialogInterface dialog,int which){
    textMessage.setText("用户已确认");
}
    };
builder.setPositiveButton("确定",listener);
AlertDialog dialog=builder.create();
Dialog.show();
}
```

这是最简单的提醒对话框,其中使用了 setTitle()设置名称,setIcon()设置显示图片,setMessage()显示信息等。通过 OnClickListener()监听按钮单击事件,使用 setPositiveButton()方法添加"确定"按钮。使用 Dialog.show()显示界面,最终显示效果如图9.11所示。

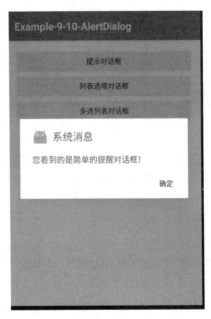

图9.11 简单的提醒对话框

在实际应用中,很多应用中不仅要有确认按钮,还要提供取消操作,常用的做法是不在对话框上显示取消按钮,而是允许用户单击硬件回退键即可引发取消操作,此时需要在程序中添加 OnCancelListener()方法。具体方法如下。

```
builder.setCancelable(true);
builder.setOnCancelListener(
```

```
new DialogInterface.OnClickListener(){
@Override
public void onCancel(DialogInterface dialog){
        textMessage.setText("用户单击回退键取消操作");
}
        }
        );
```

Android 的 alertDialog 中封装好了几个 Button，分别是 PositiveButton、NegativeButton、NeutralButton。这些 Button 和普通的 Button 没有区别，可以写任意的方法，只是命名不同。它们代表确定、否定和中立。3 个 Button 可以写任意方法，只是位置不同而已，确定 Button 一般靠左，这是阅读习惯。这些按钮仍然可以在 Activity 中使用，将例 9-10 修改如下。

```
builder.setPositiveButton("",new DialogInterface.OnClickListener(){
@Override
public void onClick(DialogInterface dialog,int which){
        textMessage.setText("用户单击 Positive 操作");
}
        });
builder.setNegetiveButton("",new DialogInterface.OnClickListener(){
@Override
public void onClick(DialogInterface dialog,int which){
        textMessage.setText("用户单击 Negetive 操作");
}
        });
builder.setNeutralButton("",new DialogInterface.OnClickListener(){
@Override
public void onClick(DialogInterface dialog,int which){
        textMessage.setText("用户单击 Neutral 操作");
}
        });
```

显示效果如图 9.12 所示。

除了上述简单的提醒对话框外，经常使用的还有列表选项对话框及多选按钮对话框，列表选项对话框的监听使用的是 setItems 方法。在例 9-10 的布局文件中添加一个 id 为 button1 的按钮，单击该按钮时实现弹出列表选项对话框，在其对应的 Activity 中实现的代码如下。

```
button1=(Button)findViewById(R.id.button1);
button1.setOnClickListener(new View.OnClickListener(){
        @Override
        public void onClick(View view){
            AlertDialog.Builder builder=new AlertDialog.Builder(MainActivity.this);
            builder.setTitle("你来自哪个系:");
```

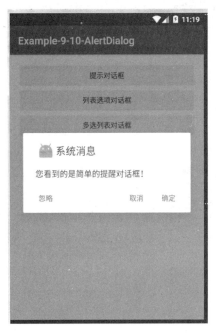

图 9.12　AlertDialog 显示 3 个按钮

```
builder.setIcon(R.mipmap.ic_launcher);
final String[] department={"软件工程系","计算机","日语","数字艺术"};
builder.setItems(department, new DialogInterface.OnClickListener(){
    @Override
    public void onClick(DialogInterface dialogInterface, int i){
        Toast.makeText(MainActivity.this,department[i],Toast.LENGTH_
        SHORT).show();
    }
});
builder.show();
    }
});
```

上述代码中使用了 builder.setItems()方法实现对列表项的监听。类似地，可以使用 setMultiChoiceItems()方法对多选按钮对话框进行设置。接下来在布局文件中定义一个 button2，单击该按钮时弹出多选按钮对话框，具体代码如下。

```
button2=(Button)findViewById(R.id.button2);
    button2.setOnClickListener(new View.OnClickListener(){
        @Override
        public void onClick(View view){
            final AlertDialog.Builder builder=new AlertDialog.Builder
            (MainActivity.this);
            builder.setTitle("你来自哪个系:");
```

```java
        builder.setIcon(R.mipmap.ic_launcher);
        final String[] department={"软件工程系","计算机","日语","数字艺术"};
        final ArrayList<String>list=new ArrayList<String>();
        builder.setMultiChoiceItems(department, null, new DialogInterface.
OnMultiChoiceClickListener(){
            @Override
            public void onClick(DialogInterface dialogInterface, int i,
            boolean b){
                if(b){
                    list.add(department[i]);
                }
                else{
                    list.remove(department[i]);
                }
            }
        });
        builder.setPositiveButton("确定", new DialogInterface.
        OnClickListener(){
            @Override
            public void onClick(DialogInterface dialogInterface, int i){
      Toast.makeText(MainActivity.this,list.toString(),Toast.LENGTH_SHORT).
show();
            }
        });
        builder.show();
    }
});
}
```

显示效果如图9.13所示。

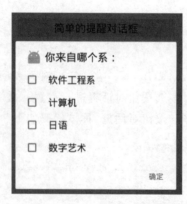

图9.13 多选按钮对话框

9.6.3 其他对话框资源

1. 日期对话框

DatePickerDialog 即包含 DatePicker 的对话框,使用该对话框需要首先创建 DatePickerDialog 对象。接下来构造函数原型 public DatePickerDialog(Context context, DatePickerDialog.OnDateSetListener callBack, int year, int monthOfYear, int dayOfMonth)。

DatePickerDialog 的主要参数如表 9.4 所示。

表 9.4 DatePickerDialog 主要参数

参数名称	参数含义
context	组件运行 Activity,DatePickerDialog.OnDateSetListener;选择日期事件
year	当前组件上显示的年
dayOfMonth	当前组件上显示的日

日期对话框与前文所述的对话框使用方法类似,使用日期对话框可以返回当前月份、日期、年份等信息。

2. 进度对话框

进度对话框通过 ProgressDialog 类实现,该类是 AlertDialog 的子类,可以直接使用 new 关键字创建。与普通对话框一样,进度条对话框最多也可以添加 3 个按钮,也可以设置风格,在默认情况下,进度条对话框是圆形进度条,如果使用水平进度条,则需要使用 setProgressStlye 方法设置。

9.7 动画资源

Android 支持两种类型的动画:渐变动画(Tweened animations)和帧动画(frame-by-frame animations)。渐变动画是对 Android 中的 View 增加渐变动画效果。帧动画是显示 drawable 目录下面一组有序图片,用于播放一个类似 GIF 的图片效果。

1. 帧动画

PC 中的动画有 Flash 和 GIF,GIF 动画是由多帧组成的动画图片格式。Android 目前不支持 GIF 动画,只能通过帧动画(frame-by-frame)实现。

【例 9-11】 帧动画实例。

使用帧动画时,首先需要在布局文件里定义该显示样式。

```
<?xml version="1.0" encoding="utf-8"?>
<LinearLayout xmlns:android="http://schemas.android.com/apk/res/android"
android:layout_width="fill_parent"
android:layout_height="fill_parent"
```

```xml
android:orientation="vertical">

<ImageView
android:id="@+id/image"
android:layout_width="wrap_content"
android:layout_height="wrap_content"
android:src="@drawable/tuzi1"></ImageView>
</LinearLayout>
```

定义好样式后,需要编制动画。通过 XML 格式的文件指定动画帧的播放顺序和延迟时间。frame_animation.xml 要放在/res/drawable 目录下面,代码如下。

```xml
<animation-list xmlns:android="http://schemas.android.com/apk/res/android"
android:oneshot="false">

<item
android:drawable="@drawable/tuzi001"
android:duration="50" />

<item
android:drawable="@drawable/tuzi002"
android:duration="50" />

<item
android:drawable="@drawable/tuzi003"
android:duration="50" />

<item
android:drawable="@drawable/tuzi004"
android:duration="50" />

<item
android:drawable="@drawable/tuzi005"
android:duration="50" />

<item
android:drawable="@drawable/tuzi006"
android:duration="50" />

<item
android:drawable="@drawable/tuzi007"
android:duration="50" />

<item
android:drawable="@drawable/tuzi008"
```

```
android:duration="50" />
</animation-list>
ImageView imgView= (ImageView)findViewById(R.id.animationImage);
imgView.setBackgroundResource(R.drawable.frame_animation);
AnimationDrawable frameAnimation= (AnimationDrawable)imgView
        .getBackground();
        if(frameAnimation.isRunning()){
        frameAnimation.stop();
}else{
        frameAnimation.stop();
frameAnimation.start();
}
```

例9-11代码中使用了AnimationDrawable对象。
AnimationDrawable可以通过调用start()和stop()方法控制动画开始和停止。
AnimationDrawable可以通过下面的方法获得对象。

```
imgView.setBackgroundResource(R.drawable.frame_animation);
AnimationDrawable frameAnimation= (AnimationDrawable)imgView.getBackground();
```

2. 补间动画

补间动画包括移动补间TranslateTween、缩放补间ScaleTween、旋转补间Rotate、透明度补间Alapha,左右滑动效果等。

每个应用上面都可以看到左右滑动效果,不管是微博还是QQ等。实现左右滑动的方式很多,会用到的技术有ViewFlipper、GestureDetector和Animation。以下三个类起主要作用：ImageSwitcher、TextSwitcher和ViewFlipper。ImageSwitcher用来切换ImageView,TextSwitcher用来切换TextView。ViewFlipper可以在任意View之间切换。

实现关系如图9.14所示。

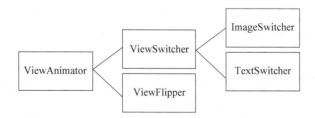

图9.14 ImageSwitcher、ImageView和TextSwitcher的关系

初始界面的实现代码如下。

```
for(int I=0;I<imgs.length;i++){
        ImageView iv=new ImageView(this);
iv.setImageResource(imgs[i]);
```

```
iv.setScaleType(ImageView.ScaleType.FIT_XY);
viewFlipper.addView(iv);
}
```

这里重点讲解 ImageView 的属性 android:scaleType,即 ImageView. setScaleType (ImageView. ScaleType)。android:scaleType 是控制图片裁剪/移动来匹配 ImageView 的 size。FIT_XY / fitXY 把图片不按比例扩大/缩小到 View 的大小显示。

9.8 风格资源与主题

在 Web 开发中,HTML 负责内容部分,CSS 负责具体实现。在 Android 开发中同样也可使用 Theme、Style+UI 组件的方式实现内容和形式的分离,做到界面的自定义。

- Style(风格、样式)是一个包含一种或多种属性的集合,可以应用到一系列 UI 组件上。这几种组件风格相似,从而实现对 UI 组件的美化。
- Theme(主题)也是一个包含一种或多种格式化属性的集合。和风格不同,主题可以应用在整个应用程序或某个窗口中。

主题和风格的用途相似,一般风格用于一个 UI 组件的美化,一个界面可以有一种或多种风格。而主题不对单个组件美化,而是对界面或整个程序进行美化。下面分别介绍主题资源和样式资源。

9.8.1 风格资源

风格(Style)资源可以看做将一系列控件的属性统一到一起,成为一种可重用的资源。声明在 res/values/styles.xml 文件中,并且要为每一个 style 定义一个名称。使用时,在布局文件中对控件使用 style="@styles/<style 的名称>"进行引用,在控件中定义的属性可以覆盖 style 中对应的属性。

【例 9-12】 Style 资源的使用。

本程序为按钮定义显示样式,在 values/styles.xml 文件中定义样式的代码如下。

```
<resources>
    <style name="my_button_style">
        <item name="android:layout_width">fill_parent</item>
        <item name="android:layout_height">wrap_content</item>
        <item name="android:textSize">30sp</item>
    </style>
</resources>
```

以上代码使用了<item></item>标记定义了具体样式,样式名称设置为 my_button_style,字体大小为 30sp。下面在布局文件中加入一个按钮,对该按钮使用上述样式。代码如下。

```
<Button
    style="@style/my_button_style"
```

```
        android:text="Button A" />

    <Button
        style="@style/my_button_style"
        android:text="Button B" />
```

以上代码为 Button A 和 Button B 设置了显示样式。Android 中还支持继承样式的功能，只要在<style></style>标记中使用 parent 属性设置即可。

将例 9-12 中的样式文件作如下修改。

```
<style name="my_button_style">
    <item name="android:layout_width">fill_parent</item>
    <item name="android:layout_height">wrap_content</item>
    <item name="android:textSize">30sp</item>
</style>

<style name="my_button_sub_style" parent="my_button_style">
    <item name="android:textColor">#F00</item>
</style>
```

此时，my_button_sub_style 就继承了 my_button_style 属性，即字号为 30ps。在布局文件中对 B 按钮使用 my_button_sub_style 样式，代码如下。

```
<Button
    style="@style/my_button_style"
    android:text="Button A" />

<Button
    style="@style/my_button_sub_style"
    android:text="Button B" />
```

由于设置了字体颜色为红色，my_button_sub_style 样式又具有红色字体的特点，所以显示效果如图 9.15 所示（此时 BUTTON B 显示为红色）。

如果在 my_button_sub_style 中也设置一个属性值，比如<item name="android:textSize">10sp</item>，此时会使用字样式的值，即 10sp。

【例 9-13】 控件圆角矩形效果。

下面为控件设置圆角矩形效果。控件只要有 background 属性，就可以引用 XML 文件。首先新建资源文件，本例中在 res/drawable 文件下添加名为 shape 的 XML 文件。主要方法为在 drawable 下右击，在弹出的快捷菜单中选择 new/Drawable resource file，将 File name 命名为 shape，添加代码如下。

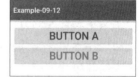

图 9.15　对按钮使用 style

```
<?xml version="1.0" encoding="utf-8"?>
<shape xmlns:android="http://schemas.android.com/apk/res/android">
    <corners android:radius="8dp"></corners>
```

```
    <gradient
        android:angle="45"
        android:endColor="#0000FF"
        android:startColor="#FFFF0000"></gradient>
</shape>
```

其中 corners 的作用是控制圆角的半径,值越大半径越大,圆角的弧度就越大。在布局文件中引用上述圆角矩形效果,代码如下。

```
<Button
    style="@style/my_button_style"
    android:text="Button A" />

<Button
    style="@style/my_button_sub_style"
    android:text="Button B"
    android:background="@drawable/shape"/>
```

最终显示效果如图 9.16 所示。

图 9.16　控件圆角矩形效果

9.8.2　主题资源

主题(Theme)资源是一系列资源的一种特殊样式,可以更换应用程序或某个 Activity 的主题,以改变主题来改变应用程序运行的外观。Android 系统已经提供了一系列主题资源,在 Android 开发过程中,可以使用继承某个主题并覆盖其中某些项的方法来定义自己的主题资源。

定义主题资源的方法是在资源文件里建立不同样式的主题风格,使用时通过操作者的不同操作来切换主题资源。

下面介绍一个定义主题资源的例子,单击该按钮显示不同主题效果,如图 9.17 所示。

图 9.17　在 Android 中设置不同主题

【例 9-14】　定义及使用主题资源。

首先设置两个不同的主题风格,一个是黑色主题,一个是白色主题。分别设置字号和字体显示颜色，style 代码如下。

```
<style name="Theme_a" parent="Theme.AppCompat">
    <item name="android:textSize">30sp</item>
    <item name="android:textColor">#F00</item>
    <item name="android:background">#FF000000</item>
</style>

<style name="Theme_b" parent="Theme.AppCompat.Light">
    <item name="android:textSize">30sp</item>
    <item name="android:textColor">#F00</item>
    <item name="android:background">#15d1ae</item>
</style>
```

使用 Theme 风格时需要导入 V7 中的 appcompat LIB 库工程，编译后才能引用，否则会产生 You need to use a Theme.AppCompat theme(or descendant)with this activity 错误。

在主布局文件中定义按钮控件，布局文件代码如下。

```
<Button
    android:layout_width="wrap_content"
    android:layout_height="wrap_content"
    android:id="@+id/btn"
    android:text="单击该按钮更换主题" />
```

对程序使用 Theme_b 主题，需要更改 Manifest 对应的 Theme，关键代码如下。

```
<application
    android:allowBackup="true"
    android:icon="@mipmap/ic_launcher"
    android:label="@string/app_name"
    android:supportsRtl="true"
    android:theme="@style/Theme_b">
    <activity android:name=".MainActivity">
        <intent-filter>
            <action android:name="android.intent.action.MAIN" />
            <category android:name="android.intent.category.LAUNCHER" />
        </intent-filter>
    </activity>
</application>
```

单击按钮后更换为 Theme_a 主题，此时需要在 Activity 中实现 OnClick 事件，以更换对应的主题，具体代码如下。

```
private Button btn;
@Override
```

```java
protected void onCreate(Bundle savedInstanceState){
    super.onCreate(savedInstanceState);
    setContentView(R.layout.activity_main);
    btn=(Button)findViewById(R.id.btn);
    btn.setOnClickListener(new View.OnClickListener(){
        @Override
        public void onClick(View v){
            setTheme(R.style.Theme_a);
            setContentView(R.layout.activity_main);
        }
    });
}
```

9.8.3 图像状态资源

Android SDK 提供的 Button 控件默认样式有些单调，为此，Android 提供了一种改变 Button 默认样式的方法，该方法不需要编写一行 Java 代码。

按钮处于不同状态（正常、按下、获得焦点等）时显示不同样式，这些样式一般使用不同的图像来渲染。这就需要指定与不同状态对应的图像，而图像状态（State）资源就是用来指定这些图像的。

图像状态资源是 XML 格式的文件，必须以＜selector＞标签作为根节点。＜selector＞标签中包含若干个＜item＞标签，用来指定相应的图像资源。

【例 9-15】 修改 Button 样式。

假设 3 个图像 normal.png、focusd.png 和 pressed.png 分别表示按钮默认的样式、获得焦点的样式和被按下的样式。在 res/drawable 目录建立一个 button.xml 文件，并输入如下内容。

```xml
<?xml version="1.0" encoding="utf-8"?>
<selector xmlns:android="http://schemas.android.com/apk/res/android">
    <item android:state_pressed="true"
        android:drawable="@drawable/pressed" /><!--pressed-->
    <item android:state_focused="true"
        android:drawable="@drawable/focused" />
    <item android:drawable="@drawable/normal" />
</selector>
```

程序分析：＜selector＞标签中有 3 个＜item＞标签，前两个＜item＞标签分别将 android:state_pressed 和 android:state_foucsed 的属性值设为 true，后一个表示当前＜item＞标签的 android:drawable 属性指定的图像是被按下和获得焦点的样式。

在布局文件中定义一个＜Button＞标签，并按以下代码设置＜Button＞标签的属性值。

```xml
<Button
```

```
android:layout_width="wrap_content"
android:layout_height="wrap_content"
android:background="@drawable/button"
android:text="按钮"/>
```

运行程序会显示图 9.18(a)所示的默认样式。按下这个按钮(不要抬起来),会显示图 9.18(b)所示的按钮按下后的样式。

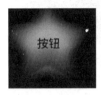

(a) 按钮样式　　　　(b) 按钮按下的样式

图 9.18　修改 Button 的默认样式效果图

9.9　国际化(I18N)

国际化的英文拼写为 Internationalization,简化后一般称为 I18N。随着 Android 的快速发展,应用程序的国际化需求也日益增加——就要求开发者开发应用程序时考虑不同的使用人群。比如设置手机语言时,程序能根据用户选择语言,加载相对应的文件。对用户来说,感受到的是程序的本地化,而对于开发人员来说,就是程序实现了国际化。

比如,开发者需要设置程序显示的字体资源,此时就不单单在资源文件夹内设置 strings 文件,还需要设置对应的语言文件。比如需要显示繁体中文,则需建立 Values-zh-rTW 文件夹,内部同样建立一个 strings 文件并进行对应设置,常见的语言设置参见表 9.5。

表 9.5　字符串资源国际化命名规则

参　数　名　称	描　　　　述
Values/strings.xml	默认使用
Values-zh-rCN/strings.xml	中文(简体)
Values-zh-rTW/strings.xml	中文(繁体)
Values-en-rUS/strings.xm	英文

如图 9.19 所示,在开发环境中,开发者需要切换到 Project 结构,在该结构下新建资源文件夹,并进行对应的设置。

【例 9-16】　字符串资源的国际化。

如图 9-16 所示,建立 Values-en-rUS 文件,在该文件下新建并设置对应的 strings 文件,代码如下。

```
<resources>
```

图 9.19　Project 结构下的值资源目录

```
    <string name="app_name">Example-09-18</string>
    <string name="btn1">Button1</string>
    <string name="btn2">Button2</string>
    <string name="btn3">Button3</string>
    <string name="btn4">Button4</string>
</resources>
```

同时默认的 strings 文件代码如下。

```
<resources>
    <string name="app_name">Example-09-18</string>
    <string name="btn1">第一个按钮</string>
    <string name="btn2">第二个按钮</string>
    <string name="btn3">第三个按钮</string>
    <string name="btn4">第四个按钮</string>
</resources>
```

布局文件使用竖排排列的线性布局，代码如下。

```
<Button
    style="@style/my_button_style"
    android:text="@string/btn1"/>
<Button
    style="@style/my_button_style"
    android:text="@string/btn2"/>
<Button
    style="@style/my_button_style"
    android:text="@string/btn3"/>
<Button
    style="@style/my_button_style"
    android:text="@string/btn4"/>
```

最终程序显示结果如图 9.20 所示。

图 9.20　默认的按钮显示结果

调整模拟器的字体，则程序自动转换显示效果，如选择模拟器的默认字体为英文时，程序显示结果如图 9.21 所示。

BUTTON1
BUTTON2
BUTTON3
BUTTON4

图 9.21　英文字体下程序的按钮显示结果

总之，使用程序时会自动根据系统语言来显示对应的信息。如果开发者没有设置对应该语言的字符串资源，程序则会显示 Values/strings.xml 中默认使用的资源。

9.10　项目实战：CoffeeStore 中各种资源的使用

9.10.1　项目分析

项目中需要使用各种不同的资源。如图 9.22 所示，Activity 里会使用大量图片资源（app/res/drawable），每个 Activity 的页面会使用布局资源（app/res/layout）。为了更好地动态显示项目效果，也会用到动画资源（app/res/anim）。值资源（app/res/value）也会频繁地使用，使用较多的值资源包括属性、颜色、样式、字符串等。

图 9.22　项目中的资源文件

9.10.2　项目实现

1. 字符串资源的使用

在本项目中,字符串资源被大量使用。具体方法是在 strings.xml 文件中声明不同位置的字符串资源并使用。如果想修改某一位置的字符串信息,不需要在程序文件中寻找,而是直接在字符串资源文件中修改。这样大大简化了修改难度,如果想将程序国际化,直接将字符串资源中的文字替换成需要的文字即可。在本项目中,具体字符串资源的使用代码如下。

```xml
<?xml version="1.0" encoding="utf-8"?>
<resources>
    <string name="login_name">登录</string>
    <string name="welcome_to_coffeeonline_system">欢迎来到在线咖啡销售系统</string>
    <string name="app_name">CoffeeStore</string>
    <string name="hello_world">Hello world!</string>
    <string name="action_settings">Settings</string>
    <string name="title_activity_main">MainActivity</string>
    <string name="hello_blank_fragment">Hello blank fragment</string>
    ……
    <string name="title_activity_test_main">TestMainActivity</string>
    <string name="title_activity_update_shop">UpdateShopActivity</string>
    <string name="title_activity_add_shop">AddShopActivity</string>

</resources>
```

上述的字符串资源在 strings.xml 文件中定义之后,使用时可以通过加载对应布局文件并寻找对应名字的方式获取。它实现了所有 Strings 资源的统一管理,而且更加有利于搜索使用及修改对应的字符串资源。

2. 图像状态资源

前面使用了单选按钮作为底部导航条的案例,但是并没有添加样式。本节将使用选择器资源给单选按钮加上样式以及状态,实现图 9.23 所示的比较漂亮的导航条。该导航条在未选择状态下显示的是白色字体灰色图片,当选择了某一具体项目时,字体颜色会相应地变为绿色图片且高亮,例如"我的"按钮,未选择时状态如图 9.23(a)所示,选择后如图 9.23(b)所示,这就是使用了 Selector 的效果。

(a) 主页底端导航条"我的"未选中状态　　(b) 主页底端导航条"我的"选中状态

图 9.23　主页底端导航条"我的"状态

【例 9-17】 使用资源文件美化 RadioButton 控件。

前面讲解了使用 RadioButton 控件来控制 Fragment 的方法,将 RadioButton 控件作为导航条使用,但是显示效果不好,下面使用资源文件美化该控件。

首先创建该导航条的布局文件,使用 alignParentBottom 属性,将其布局到页面底端,并对 RadioGroup 设置背景图片为 tabbar_bg,布局文件具体代码如下。

```xml
<RadioGroup
    android:id="@+id/tabbar"
    android:layout_width="fill_parent"
    android:layout_height="wrap_content"
    android:layout_alignParentBottom="true"
    android:background="@drawable/tabbar_bg"
    android:orientation="horizontal">

    <RadioButton
        android:id="@+id/tab_weather"
        style="@style/main_tab_bottom_style"
        android:drawableTop="@drawable/tab_weather_selector"
        android:text="@string/tab_text_weather" />

    <RadioButton
        android:id="@+id/tab_trend"
        style="@style/main_tab_bottom_style"
        android:text="@string/tab_text_trend" />

    <RadioButton
        android:id="@+id/tab_index"
        style="@style/main_tab_bottom_style"
        android:text="@string/tab_text_index" />

    <RadioButton
        android:id="@+id/tab_tools"
        style="@style/main_tab_bottom_style"
        android:text="@string/tab_text_tools" />

    <RadioButton
        android:id="@+id/tab_setting"
        style="@style/main_tab_bottom_style"
        android:text="@string/tab_text_setting" />

</RadioGroup>
```

下面以第一个单选按钮为例进行设置。首先设置其 id 属性为 tab_weather,同时在 strings.xml 文件下设置其显示文字为天气,之后为其设置 style 样式。具体代码如下。

```xml
<string name="tab_text_weather">首页</string>
```

```xml
<style name="main_tab_bottom_style">
    <item name="android:textColor">#faf5f5</item>
    <item name="android:textSize">11dp</item>
    <item name="android:ellipsize">marquee</item>
    <item name="android:gravity">center_horizontal</item>
    <item name="android:singleLine">true</item>
    <item name="android:button">@null</item>
    <item name="android:layout_width">0dp</item>
    <item name="android:layout_height">wrap_content</item>
    <item name="android:layout_weight">0.5</item>
</style>
```

上述代码设置了该单选按钮的各种属性，其中<item name="android:button">@null</item>为去掉按钮样式的属性。最后设置 drawableTop 属性，该属性用来设置图片在文字上方，并使用 selector 设置其选中变化时产生的不同效果。

```xml
<?xml version="1.0" encoding="utf-8"?>
<selector xmlns:android="http://schemas.android.com/apk/res/android">
    <item android:drawable="@drawable/tab_weather" android:state_checked="false" />
    <item android:drawable="@drawable/tab_weather_checked" android:state_checked="true" />
</selector>
```

下面设置选择按钮时图片颜色由灰色变为高亮，方法是在 drawable 文件夹下的 icon_*.xml 文件中修改。本项目中共有首页、购物车、分类、搜索、和"我的"等 5 个可选择按钮。分别设置这 5 个可选框的 XML 文件，下面以 icon_me.xml 为例修改代码如下（其他几个的设置方法与这个相似）。

```xml
<?xml version="1.0" encoding="utf-8"?>
<selector xmlns:android="http://schemas.android.com/apk/res/android">
    <item android:drawable="@drawable/icon_me_nor" android:state_checked="false" />
    <item android:drawable="@drawable/icon_me_sel" android:state_checked="true" />
</selector>
```

设置完成后，需要在其对应的 Java 文件中引用该布局文件，代码如下。

```java
public void InitView(){
    main_tab_RadioGroup= (RadioGroup)findViewById(R.id.main_tab_RadioGroup);

    radio_home= (RadioButton)findViewById(R.id.radio_home);
    radio_shopcar= (RadioButton)findViewById(R.id.radio_shopcar);
    radio_sort= (RadioButton)findViewById(R.id.radio_sort);
    radio_search= (RadioButton)findViewById(R.id.radio_search);
```

```java
        radio_me=(RadioButton)findViewById(R.id.radio_me);

        main_tab_RadioGroup.setOnCheckedChangeListener(this);
    }

    public void InitViewPager(){
        main_viewPager=(ViewPager)findViewById(R.id.main_ViewPager);

        fragmentList=new ArrayList<Fragment>();

        Fragment homeFragment=new HomeFragment();
        Fragment sortFragment=new SortFragment();
        Fragment shopCarFragment=new ShopCarFragment();
        Fragment searchFragment=new SearchFragment();
        Fragment meFragment=new MeFragment();

        fragmentList.add(homeFragment);
        fragmentList.add(shopCarFragment);
        fragmentList.add(sortFragment);
        fragmentList.add(searchFragment);
        fragmentList.add(meFragment);
        //设置 ViewPager 适配器
        main_viewPager.setAdapter(new MyAdapter(getSupportFragmentManager(),
        fragmentList));
        main_viewPager.setCurrentItem(0);
        main_viewPager.setOnPageChangeListener(new MyListner());
    }
    public class MyAdapter extends FragmentPagerAdapter {
        ArrayList<Fragment>list;
        public MyAdapter(FragmentManager fm, ArrayList<Fragment>list){
            super(fm);
            this.list=list;
        }
        @Override
        public Fragment getItem(int arg0){
            return list.get(arg0);
        }

        @Override
        public int getCount(){
            return list.size();
        }
    }
```

项目中使用动画的场景很多，比如首页的图片切换，目标被选中时样式的改变等内容,具体定义的动画资源如图 9.24 所示。以 slide_in_right.xml 文件为例介绍动画的显示过程。

首先在 slide_in_right.xml 文件中定义动画进入的位置(fromXDelta)、最终的动画位置(toXDelta)以及动画持续时间。其中 fromXDelta 属性为动画起始时 X 坐标上的位置;toXDelta 属性为动画结束时 X 坐标上的位置;fromYDelta 属性为动画起始时 Y 坐标上的位置;toYDelta 属性为动画结束时 Y 坐标上的位置。值得注意的是,没有指定的属性默认是以自己为相对参照物,这里没有指定 Y 轴的属性,则默认为其自身在 Y 轴方向没变化。另外,文件同时也定义了动画进入时透明度的变化,0.0 表示完全透明,而 1.0 表示完全不透明,进入效果是一个从无到有的过程。另外,duration 属性为动画持续时间,时间以毫秒为单位。

图 9.24　三级项目中的动画资源

```xml
<?xml version="1.0" encoding="utf-8"?>
<set xmlns:android="http://schemas.android.com/apk/res/android">
<translate
android:duration="300"
android:fromXDelta="50%p"
android:toXDelta="0" />
<alpha
android:duration="300"
android:fromAlpha="0.0"
android:toAlpha="1.0" />
</set>
```

9.10.3　项目说明

本节讲解各种资源的使用及资源在实际项目中的应用方式,其中字符串资源是使用最多的资源,合理使用字符串资源将使页面显示更加多样化,对后期的国际化有很大好处。定义好尺寸资源、颜色和数组资源后,可以在项目中随时调用,使编程更加灵活。通过菜单资源实现页面隐藏更多的信息,丰富了用户的交互形式。对话框资源提供更多的提示信息,动画资源使页面显示更加多样化,风格和主题资源使程序统一显示。总之,资源文件是 Android 开发中必不可少的部分,知识相对容易,但是合理运用可以实现多样化的效果,这需要开发者深入钻研,多多实践。

本章小结

本章全面系统地介绍了 Android 程序开发中需要的各种资源文件。读者应重点掌握不同资源文件存放的位置、资源的属性、调用的方法等内容,熟练掌握并使用字符串资源、图片资源、值资源、风格与主题资源、动画资源等知识。学好使用资源文件,可以更好地适应越来越发达的网络环境,适应程序国际化、本地化的必然趋势。

本章习题

一、选择题

1. 在下面的文件(文件夹)中,放置图片资源的一项是(　　)。
   A. res/values/   B. /values/strings.xm
   C. /res/values/styles.xml   D. /app/res/drawable/
2. 下面哪一项不是继承自 AlertDialog 的?(　　)
   A. ProgressDialog   B. DatePickerDialog
   C. ActionBar   D. TimePickerDialog
3. Android 中不属于显示单位的是(　　)。
   A. px   B. dip   C. dpi   D. sp
4. 下面哪一项不是 Android 中的菜单类型?(　　)
   A. DateMenu   B. SubMenu   C. ContextMenu   D. OptionsMenu
5. 下列说法中错误的一项是(　　)。
   A. Toast(提示)是 Android 中用来显示提示信息的一种机制
   B. Notification(通知)是 Android 提供的出现在状态栏的提醒机制
   C. Toast 没有焦点且显示的时间有限,不会打断用户当前的操作,不能与用户交互
   D. 用户打开 Notification 之后,会显示通知信息,不可与用户进行交互,也不能处理用户选择事件

二、简答题

1. 写出创建一个菜单的步骤。
2. 写出主题与样式的区别。
3. 写出自定义样式的步骤。
4. 写出继承样式的两种方式。
5. 说明 CoffeeStore 项目中资源文件的使用位置,并讨论还可以如何使用资源文件。

Android人机交互设计

本章概述

前面章节讲解了各控件与资源的使用,但对用户来说,更关注的是界面美观程度和操作使用程序的方法,这就涉及本章所讲述的Android人机交互知识。

在Android程序中,用户主要通过单击、滑动或手势控制的方式进行操作,反映到程序中就是对事件的处理。将控件与事件处理相结合,就会实现应用程序的基本外部骨架。

本章主要讲述常用事件,包括按键事件和触摸事件,实现拖拉与多点触屏,并且在此基础上实现手势识别等交互性设计。

学习重点与难点

重点:

(1) 触屏事件。

(2) 拖拉与多点触屏。

(3) 手势识别。

难点:

(1) 多点触屏。

(2) 手势识别的设定与实现。

学习建议

本章的人机交互知识涵盖了很多接口,包括OnClickListener、OnLongClickListener、OnKeyListener、OnKeyDown等,这些接口是程序开发中常用的知识,应用时往往都是对这些接口进行扩展,使程序具有更好的人机交互体验,评判应用程序是否优秀的标准往往也是看用户体验度是否良好。

加强对常用事件、手势识别等技术的理解和掌握,在开发中学会灵活运用各种不同的交互效果,可以提高应用程序的用户体验度。

10.1 常用事件

目前的图形界面应用程序都是通过事件实现人机交互的。事件可以理解为用户对图形界面的操作。同样,Android的界面框架也支持对按键事件的监听,并能够将按键事件的详细信息传递给处理函数。程序监听事件,并对监听到的事件进行响应,从而实现与用户之间的交互。

本章介绍 Android 中的常用事件,包括按键事件、拖拉与多点触屏及手势识别。

10.1.1 按键事件

View 组件将事件分为几种,如表 10.1 所示。

表 10.1　View 组件事件接口表

接　　口	描　　述	回 调 方 法
OnClickListener 接口	单击事件	public void onClick(View v)v 为单击发生事件的组件
OnLongClickListener 接口	长按事件	public boolean onLongClick(View v)返回 true,表示事件已经处理完毕
OnFocusChangeListener 接口	组件焦点改变事件	public void onFocusChange(View v, boolean hasFocus);hasFocus 表示事件源的状态,是否获得焦点
OnKeyListener 接口	手机键盘事件的监听	public boolean onKey(View v, int keyCode, KeyEvent event)keyCode 为键盘码,event 为键盘事件封装的对象
OnTouchListener 接口	处理手机屏幕事件	public boolean onTouch(View v,MotionEvent)触摸、按下、抬起、滑动都会触发该事件
OnCreateContextMenuListener 接口	处理上下文菜单被创建的事件	

【例 10-1】 对控件设置长按事件。

设置单击事件会使用 OnLongClickListener 接口,该接口为 View 类长按事件的捕捉接口,即当长时间按某个 View 类对象或其子类对象时触发的事件。返回值是一个 boolean 型变量,返回 true 时表示已经完成,返回 false 时表示没有完全处理完该事件。下面对 EditView 控件设置长按事件。

首先在布局文件中定义 EditText 控件,关键代码如下。

```
<TextView
    android:id="@+id/textView"
    android:layout_width="wrap_content"
    android:layout_height="wrap_content"
    android:text="下面按钮设置了长按事件" />

<EditText
    android:id="@+id/editText"
    android:layout_width="wrap_content"
    android:layout_height="wrap_content"
    android:layout_alignParentLeft="true"
    android:layout_alignParentStart="true"
    android:layout_below="@+id/textView"
```

```
android:ems="10" />
```

同时,在其对应的 Activity 中引用该控件,并对其使用长按事件,设置效果为长按该控件时弹出"您在长按此控件"。首先定义 EditText 为 et1,并通过 findViewById 获得该控件,之后对该控件使用 OnLongClickListener 接口,长按时通过 Toast 弹出需要显示的文字。代码如下。

```
package cnt.edu.neusoft.software.mobile.example_10_1;

import android.os.Bundle;
import android.support.v7.app.AppCompatActivity;
import android.view.View;
import android.widget.EditText;
import android.widget.Toast;
public class MainActivity extends AppCompatActivity {
    @Override
    protected void onCreate(Bundle savedInstanceState){
        super.onCreate(savedInstanceState);
        setContentView(R.layout.activity_main);
        EditText et1=(EditText)findViewById(R.id.editText);
        et1.setOnLongClickListener(new View.OnLongClickListener(){
            @Override
            public boolean onLongClick(View v){
                Toast.makeText(MainActivity.this, "您在长按此控件", Toast.LENGTH
                _SHORT).show();
                return false;
            }
        });
    }
}
```

例 10-1 实现了长按事件,可以使用同样方法实现单击事件。方法与上述类似,只需要将接口替换为 OnClickListener 接口,下面将例 10-1 进行扩展,实现手机键盘事件的监听。

【例 10-2】 设置手机键盘事件的监听。

本例的目的是设置手机键盘事件监听,当使用键盘输入信息时,将其在屏幕中显示出来。主要原理是对该 EditText 控件实现 OnKeyListener 接口,实现对接口中的按键码进行监控,最后将获得的输入数据通过 Toast 在界面中显示出来。

这里介绍按键事件的几个参数,如表 10.2 所示。

表 10.2 按键事件的几个参数

位 置	内 容 描 述
第 1 个参数 view	表示产生按键事件的界面控件
第 2 个参数 keyCode	表示按键代码
第 3 个参数 keyEvent	包含事件的详细信息,如按键的重复次数、硬件编码和按键标志等

其中,XML 文件与例 10-1 的布局文件相同,而在 Activity 中获取 OnKeyListener 接口的 keyCode 参数值,并通过 Toast 在界面中显示出来。

Activity 中的关键代码如下:

```
et1.setOnKeyListener(new View.OnKeyListener(){
    @Override
    public boolean onKey(View v, int keyCode, KeyEvent event){
        Toast.makeText(MainActivity.this, keyCode+"", Toast.LENGTH_SHORT).show();
        return false;
    }
});
```

【例 10-3】 设置对物理按键监听。

Android 程序中不仅可以实现对控件的监听,也可以实现对物理按键的监听。一般物理按键的具体功能在 Android 框架下已经实现,但是可以通过程序重写物理按键的方法,实现想要的功能,这样开发的程序更加丰富,更加多样化。下面使用 onKeyDown 方法监听物理键的返回键(KEYCODE_BACK),重写返回事件。代码如下。

```
public class MainActivity extends AppCompatActivity {

    long firstClickTime;
    @Override
    protected void onCreate(Bundle savedInstanceState){
        super.onCreate(savedInstanceState);
        setContentView(R.layout.activity_main);
        EditText et1=(EditText)findViewById(R.id.editText);
        et1.setOnLongClickListener(new View.OnLongClickListener(){
            @Override
            public boolean onLongClick(View v){
              Toast.makeText(MainActivity.this,"您在长按",Toast.LENGTH_SHORT).
            show();
                return false;
            }
        });
        et1.setOnKeyListener(new View.OnKeyListener(){
            @Override
            public boolean onKey(View v, int keyCode, KeyEvent event){
              Toast.makeText(MainActivity.this, keyCode+"", Toast.LENGTH_SHORT).
            show();
                return false;
            }
        });
        Button btn=(Button)findViewById(R.id.button);
        btn.setOnClickListener(new View.OnClickListener(){
```

```
            @Override
            public void onClick(View v){
                Intent intent=new Intent(MainActivity.this,Main2Activity.class);
                MainActivity.this.startActivity(intent);
            }
        });
        Button btn2= (Button)findViewById(R.id.button2);
        btn2.setOnClickListener(new View.OnClickListener(){
            @Override
            public void onClick(View v){
                Intent intent=new Intent(MainActivity.this,
                ChangeAlphaActivity.class);
                MainActivity.this.startActivity(intent);
            }
        });
    }
    @Override
    public boolean onKeyDown(int keyCode, KeyEvent event){
        if(keyCode==KeyEvent.KEYCODE_BACK){
            if(System.currentTimeMillis()-firstClickTime >=2000){
                Toast.makeText(this,"再按一次退出程序",Toast.LENGTH_SHORT).show();
                firstClickTime=System.currentTimeMillis();
            } else
                this.finish();
        }
        return false;
    }
}
```

上面接口中对 KeyEvent.KEYCODE_BACK 是否被单击进行判断，如果单击了该按键，对按键间隔时间进行判断，根据判断结果返回，关闭当前 Activity 或者弹出提示信息。

10.1.2 触摸事件

Android 界面框架支持对触摸事件的监听，并能够将触摸事件的详细信息传递给处理函数。此时需要设置触摸事件的监听器，并重载 onTouch()函数。首先设置控件的触摸事件监听器，主要代码为 touchView.setOnTouchListener(new View.OnTouchListener())。

接下来使用 public boolean onTouch(View v，MotionEvent event)重载 onTouch 函数。其中第 1 个参数 View 表示产生触摸事件的界面控件；第 2 个参数 MontionEvent 表示触摸事件的详细信息，如产生时间、坐标和触点压力等。最后设置 onTouch()函数的返回值：返回 true/false。

【例 10-4】 处理触摸事件。

图 10.1 所示为 TouchEventDemo 用户界面示例，其中浅色区域是可以接受触摸事

件的区域,用户可以在 Android 模拟器中使用鼠标单击屏幕,以模拟触摸手机屏幕。下方深色区域是显示区域,用来显示触摸事件的类型、相对坐标、绝对坐标、触点压力、触点尺寸和历史数据量等信息。

图 10.1 触摸事件示例

当手指接触到触摸屏、在触摸屏上移动或离开触摸屏时,分别会引发 ACTION_DOWN、ACTION_UP 和 ACTION_MOVE 触摸事件。而无论是哪种触摸事件,都会调用 onTouch()函数处理。事件类型包含在 onTouch()函数的 MotionEvent 参数中,可以通过 getAction()函数获取触摸事件的类型,然后根据触摸事件的不同类型进行不同处理。

关键代码如下。

```
touchView.setOnTouchListener(new View.OnTouchListener(){
  @Override
  public boolean onTouch(View v, MotionEvent event){
    int action=event.getAction();
    switch(action){
    case(MotionEvent.ACTION_DOWN):
      Display("ACTION_DOWN", event);
      break;
    case(MotionEvent.ACTION_UP):
      int historySize=ProcessHistory(event);
      historyView.setText("ACTION_UP "+historySize);
      Display("ACTION_UP", event);
      break;
    case(MotionEvent.ACTION_MOVE):
      Display("ACTION_MOVE", event);
```

```
        break;
    }
    return true;
}
});
```

为了能够使屏幕最上方的 TextView 处理触摸事件,需要使用 setOnTouchListener() 函数设置触摸事件监听器,并在 onTouch() 函数中添加触摸事件的处理过程。

MotionEvent 的参数中不仅有触摸事件的类型信息,还有触点的坐标信息。获取方法是使用 getX() 和 getY() 函数,这两个函数获取到的是触点相对于父界面元素的坐标信息。如果需要获取绝对坐标信息,则可使用 getRawX() 和 getRawY() 函数。

一般情况下,如果用户将手指放在触摸屏上,但不移动,然后抬起手指,应先后产生 ACTION_DOWN 和 ACTION_UP 两个触摸事件。如果用户在屏幕上移动手指,然后再抬起手指,则会产生事件序列:ACTION_DOWN→ACTION_MOVE→ACTION_MOVE→ACTION_MOVE→……→ACTION_UP,下面设置显示坐标的方法,具体代码如下。

```
private void Display(String eventType, MotionEvent event){
    int x=(int)event.getX();
    int y=(int)event.getY();
    float pressure=event.getPressure();
    float size=event.getSize();
    int RawX=(int)event.getRawX();
    int RawY=(int)event.getRawY();

    String msg="";
    msg+="历史数据量"+eventType+"\n";
    msg+="相对坐标"+String.valueOf(x)+","+String.valueOf(y)+"\n";
    msg+="绝对坐标"+String.valueOf(RawX)+","+String.valueOf(RawY)
        +"\n";
    msg+="触点压力"+String.valueOf(pressure)+"   ";
    msg+="触点尺寸"+String.valueOf(size)+"\n";
    labelView.setText(msg);
}
private int ProcessHistory(MotionEvent event){
    int historySize=event.getHistorySize();
    for(int i=0; i<historySize; i++){
        long time=event.getHistoricalEventTime(i);
        float pressure=event.getHistoricalPressure(i);
        float x=event.getHistoricalX(i);
        float y=event.getHistoricalY(i);
        float size=event.getHistoricalSize(i);
    }
    return historySize;
}
```

上述代码中的触点压力是一个介于 0 和 1 之间的浮点数,用来表示用户对触摸屏施加压力的大小,接近 0 表示压力较小,接近 1 表示压力较大。获取触摸事件触点压力的方式是调用 getPressure()函数的触点尺寸,它指用户接触触摸屏的接触点大小,也是一个介于 0 和 1 之间的浮点数,可以使用 getSize()函数获取。模拟器并不支持触点压力和触点尺寸的模拟,所有触点压力恒为 1.0,触点尺寸恒为 0.0。同时,模拟器上也无法产生历史数据,因此历史数据量一直显示为 0。

在手机上运行的应用程序,效率是非常重要的。如果 Android 界面框架不能产生足够多的触摸事件,则应用程序就不能很精确地描绘触摸屏上的触摸轨迹。如果 Android 界面框架产生了过多的触摸事件,虽然能够满足精度的要求,却降低了应用程序效率。Android 界面框架使用了"打包"的解决方法。触点移动速度较快时,会产生大量的数据,每经过一定的时间间隔,便会产生一个 ACTION_MOVE 事件。在这个事件中,除了有当前触点的相关信息,还包含这段时间间隔内触点轨迹的历史数据信息,这样既能够保持精度,又不致产生过多的触摸事件。

10.2 拖拉与多点触屏

多点触控技术区别于传统的单点触摸屏,最大特点是可以两只手、多个手指甚至多人同时操作屏幕,使显示更加任性。使用多点触屏时,开发者可以使用 setOnTouchListener()方法为控件设置监听器,处理监听事件。在实际应用中,多点触屏用得最多的功能就是放大缩小。比如一些图片浏览器,可以用多个手指在屏幕上操作,放大或缩小图片。再如一些浏览器,也可以通过多点触摸放大或缩小字体。其实放大缩小也只是多点触摸的实际应用样例之一,有了多点触摸技术,在一定程度上可以创新更多的操作方式,实现更酷的人机交互。

理论上说,Android 系统本身可以处理多达 256 个手指的触摸,这主要取决于手机硬件的支持程度。当然,支持多点触摸的手机也不会支持这么多点,一般支持 2 个或 4 个点。对于开发者来说,编写多点触摸的代码与编写单点触摸的代码并没有很大的差异。因为 Android SDK 中的 MotionEvent 类不仅封装了单点触摸的消息,也封装了多点触摸的消息,它们的处理方式几乎是一样的。

【例 10-5】 多点触屏监听事件,代码如下。

```
setOnTouchListener
public boolean onTouch(View arg0,MotionEvent event)
        {
switch(event.getAction()){
case MotionEvent.ACTION_DOWN:
case MotionEvent.ACTION_MOVE:
case MotionEvent.ACTION_UP:
case MotionEvent.ACTION_POINTER_DOWN:
case MotionEvent.ACTION_POINTER_UP:
        }
```

通过设置监听器的方式实现多点触控，程序采集到多点信号并判断触摸信号的意义。这样可以完成缩放、旋转等功能。

处理多点触摸的过程中，还需要用到 MotionEvent.ACTION_MASK。一般使用 switch(event.getAction()& MotionEvent.ACTION_MASK)就可以处理处理多点触摸的 ACTION_POINTER_DOWN 和 ACTION_POINTER_UP 事件。代码调用这个"与"操作以后，当第二个手指按下或放开，就会触发 ACTION_POINTER_DOWN 或 ACTION_POINTER_UP 事件。

10.3 手势识别

Android 中的很多程序都会用到手势识别，通过手指在屏幕上滑动实现各种交互效果。它类似于定义一个数据库，首先需要开发者（或用户）在数据库中定义好操作手势。之后进行操作时，程序会将手势与数据库中的手势进行匹配，并且选择最佳的匹配对象，最后实现该手势对应的功能。

手势识别的主要步骤如下。

1. 建立手势库

使用 GestureBuilder 建立手势库，创建并保存程序所需的手势。

2. 手势识别

使用 GuestOverlayView 控件来接收手势。
使用 OnGesturePerformed()方法预测手势并对比结果。

3. 结果输出

将上述所得结果与对应功能相匹配，最终导出对应功能。

【例 10-6】 下面通过案例学习手势识别的具体内容：主要实现在屏幕上识别图形手势，识别完成后显示提示信息。要实现该手势绘制并识别功能，首先需要添加 GestOverlayView 控件来接收用户手势，主要代码如下。

```
<?xml version="1.0" encoding="utf-8"?>
<LinearLayout xmlns:android="http://schemas.android.com/apk/res/android"
    android:layout_width="fill_parent"
    android:layout_height="fill_parent"
    android:background="@drawable/background"
    android:orientation="vertical">

<TextView
    android:layout_width="fill_parent"
    android:layout_height="wrap_content"
    android:gravity="center_horizontal"
```

```xml
android:text="@string/title"
android:textColor="@android:color/black"
android:textSize="20dp" />

<android.gesture.GestureOverlayView
android:id="@+id/gestures"
android:layout_width="fill_parent"
android:layout_height="0dip"
android:layout_weight="1.0" />
</LinearLayout>
```

修改 XML 文件后需要创建手势识别库，加载手势文件，接着获得布局文件中定义的 GestureOverlayView 控件。在 onGesturePerformed() 方法的实现中，获得得分最高的预测结果并提示，该代码如下。

```java
public class GesturesRecognitionActivity extends Activity implements OnGesturePerformedListener {
private GestureLibrary library;
@Override
public void onCreate(Bundle savedInstanceState){
super.onCreate(savedInstanceState);
setContentView(R.layout.main);
library=GestureLibraries.fromRawResource(this, R.raw.gestures);
                                                    //加载手势文件
if(!library.load()){                     //如果加载失败则退出
finish();
}
GestureOverlayView gesture = (GestureOverlayView) findViewById (R. id. gestures);
gesture.addOnGesturePerformedListener(this);   //增加事件监听器
}
@Override
public void onGesturePerformed ( GestureOverlayView overlay, Gesture gesture){
        ArrayList<Prediction>gestures=library.recognize(gesture);
                                              //获得全部预测结果
int index=0;                      //保存当前预测的索引号
double score=0.0;                 //保存当前预测的得分
for(int i=0; i<gestures.size(); i++){   //获得最佳匹配结果
Prediction result=gestures.get(i);      //获得一个预测结果
if(result.score>score){
            index=i;
score=result.score;
}
        }
```

```
        Toast.makeText(this, gestures.get(index).name, Toast.LENGTH_LONG).
    show();
}
```

10.4 项目实战：CoffeeStore 引导页图片切换的实现

10.4.1 项目分析

很多 Apps 第一次启动或版本更新后再启动时，都会出现几个图片切换的效果，这些图片大都是此 Apps 的功能介绍，切换后才跳到 Apps 首页。

在本项目中，当应用程序检测到首次启动时，会出现图 10.2(a) 所示的引导页面，用手指滑动界面，则出现(b)所示的欢迎界面。此启动页在项目第一次启动时才会出现。

通过手指滑动屏幕，实现从(a)到(b)的图片切换效果，这就是本章介绍的 Android 人机交互方法。

(a)　　　　　　　　　　　(b)

图 10.2　引导页的图片切换效果

10.4.2 项目实现

为实现上述图片切换效果，首先需要定义其 XML 文件，该布局文件定义名称为 viewfilter.xml。由于该布局内只有一张图片文件，所以只需要一个线性布局，该布局内部使用 ViewFlipper 控件。代码如下。

```
<?xml version="1.0" encoding="utf-8"?>
<LinearLayout xmlns:android="http://schemas.android.com/apk/res/android"
    android:layout_width="match_parent"
    android:layout_height="match_parent"
```

```xml
        android:orientation="vertical">

        <ViewFlipper
            android:id="@+id/viewflipper"
            android:layout_width="fill_parent"
            android:layout_height="fill_parent">
        </ViewFlipper>
</LinearLayout>
```

同时，在资源目录中添加动画资源文件，此动画资源用于设置手势滑动时图片进入和划出的动画效果。由于例 10-1 中布局文件比较简单，只有一个 EditText 控件，我们可以通过代码实现 OnKeyListener 接口，实现的功能是当用户输入时通过该接口获取到输入内容，并将它打印出来。具体的 Activity 中的关键代码如下。

```xml
<?xml version="1.0" encoding="utf-8"?>
<set xmlns:android="http://schemas.android.com/apk/res/android">

    <translate
        android:duration="400"
        android:fromXDelta="-100%p"
        android:toXDelta="0" />
    <alpha
        android:duration="300"
        android:fromAlpha="0.0"
        android:toAlpha="1.0" />

</set>
```

定义划入效果的同时也要定义划出时的动画效果，即对动画资源的 push_right_out.xml 文件代码修改如下。

```xml
<?xml version="1.0" encoding="utf-8"?>
<set xmlns:android="http://schemas.android.com/apk/res/android">

    <translate
        android:duration="400"
        android:fromXDelta="0"
        android:toXDelta="100%p" />
    <alpha
        android:duration="300"
        android:fromAlpha="1.0"
        android:toAlpha="0.0" />

</set>
```

在后台 Java 文件里引用该布局文件，并且定义用户手势，通过 GestureDetector 进行手势检测,检测到用户手势操作时则使用动画效果。

在手势识别中,判断手势是从左向右滑动还是从右向左滑动,要在 OnGestureListener 的 onFling 事件中根据滑动时起点与终点的坐标来判断。当检测到用户手势操作时,则使用动画效果,其代码如下。

```java
public class Welcome extends Activity implements android.view.GestureDetector.OnGestureListener{
    private int[] imgs={R.drawable.coffe1,
            R.drawable.coffee2
    };
    private GestureDetector gestureDetector=null;
    private ViewFlipper viewFlipper=null;
    private Activity mActivity=null;
    @SuppressWarnings("deprecation")
    @Override
    public void onCreate(Bundle savedInstanceState){
        super.onCreate(savedInstanceState);
        final Window win=getWindow();
        win.setFlags(WindowManager.LayoutParams.FLAG_FULLSCREEN,
        WindowManager.LayoutParams.FLAG_FULLSCREEN);
        requestWindowFeature(Window.FEATURE_NO_TITLE);
        setContentView(R.layout.viewfilter);

        mActivity=this;

        viewFlipper=(ViewFlipper)findViewById(R.id.viewflipper);
        gestureDetector=new GestureDetector(this,this);

        for(int i=0; i<imgs.length; i++){
            ImageView iv=new ImageView(this);
            iv.setImageResource(imgs[i]);
            iv.setScaleType(ImageView.ScaleType.FIT_XY);
            viewFlipper.addView(iv);
        }
    }
    @Override
    public boolean onTouchEvent(MotionEvent event){
    return gestureDetector.onTouchEvent(event);          }
    @Override
    public boolean onFling(MotionEvent e1, MotionEvent e2, float velocityX, float velocityY){
        if(e2.getX()-e1.getX()>120){
            if(viewFlipper.getDisplayedChild()==0){
                new Handler().postDelayed(new Runnable(){
```

```java
                @Override
                public void run()
                    Intent intent=new Intent(Welcome.this, MainActivity.
                    class);
                     startActivity(intent);
                     overridePendingTransition(R.anim.alpha_in, R.anim.alpha_out);
                }
            }, 1000);
        }
Animation rInAnim=AnimationUtils.loadAnimation(mActivity, R.anim.push_right_in);
Animation rOutAnim=AnimationUtils.loadAnimation(mActivity, R.anim.push_right_
out);

        viewFlipper.setInAnimation(rInAnim);
        viewFlipper.setOutAnimation(rOutAnim);
        viewFlipper.showPrevious();

    return true;
    } else if(e2.getX()-e1.getX()<-120){
        if(viewFlipper.getDisplayedChild()==0){
                    new Handler().postDelayed(new Runnable(){
            @Override
            public void run(){
                Intent intent=new Intent(Welcome.this, MainActivity.
                class);
                startActivity(intent);
                overridePendingTransition(R.anim.alpha_in, R.anim.alpha_out);
            }
        }, 1000);
        }
Animation lInAnim=AnimationUtils.loadAnimation(mActivity, R.anim.push_left_
in);
Animation lOutAnim=AnimationUtils.loadAnimation(mActivity, R.anim.push_
left_out);

        viewFlipper.setInAnimation(lInAnim);
        viewFlipper.setOutAnimation(lOutAnim);
        viewFlipper.showNext();
    }
    return true;
}

    @Override
    public boolean onDown(MotionEvent e){
```

```
        return false;
    }
    @Override
    public void onLongPress(MotionEvent e){
    }
    @Override
    public boolean onScroll(MotionEvent e1, MotionEvent e2, float distanceX,
    float distanceY)
    {
        return false;
    }
    @Override
    public void onShowPress(MotionEvent e){
    }
    @Override
    public boolean onSingleTapUp(MotionEvent e){
        return false;
    }
}
```

10.4.3 项目说明

引导页的实现使用了Android的高级控件ViewFlipper、动画资源以及手势识别等知识。手势识别是手机开发中人机交互的一种重要方式，在实际项目开发中应用场景非常多，读者应掌握这种技术。

本 章 小 结

本章介绍了手势识别、多点触碰、触摸事件等交互技术。目前滑动动画效果已经应用到各种Android程序中，它使用户更加灵活地操作手机屏幕，也使多样化的操作手机成为可能。

通过拖拉与手势识别可以实现各种自定义操作。而多点触屏技术可以实现当前所有的屏幕操作手段，也为Android程序开发指明了一个新的方向。学习Android人机交互设计，是开发复杂程序、提高用户体验效果的一个发展方向。

本 章 习 题

一、选择题

1. 当手指接触到触摸屏，在触摸屏上移动或离开屏幕时，不会产生的事件是（　　）。
   A. ACTION_DOWN   B. ACTION_UP
   C. ACTION_STOP   D. ACTION_MOVE

2. 手势识别主要步骤不包括下面哪一项？（　　）

　　A. 建立手势库　　　B. 手势识别　　　C. 手势接收　　　D. 结果输出

3. Android 中的主要人机交互事件不包括（　　）。

　　A. 多点触碰　　　B. 手势识别　　　C. 手势定义　　　D. 触摸事件

4. 关于 onTouch 函数，下面说法不正确的是（　　）。

　　A. 可以使用 public boolean onTouch（View v，MotionEvent event）重载 onTouch 函数

　　B. 第 1 个参数 View 表示产生触摸事件的界面控件；第 2 个参数 MontionEvent 表示触摸事件的详细信息，如产生时间、坐标和触点压力等

　　C. onTouch 函数没有返回值

　　D. 可以通过 getAction() 函数获取到触摸事件的类型

二、简答题

1. 写出手势识别的主要步骤及方法。
2. 写出 onClickListener 与 onLongClickListener 的区别。
3. 分析如下代码，试分析说明该代码的主要功能。

```
public boolean onKeyDown(int keyCode, KeyEvent event){
    if(keyCode ==KeyEvent.KEYCODE_BACK){
        if(System.currentTimeMillis()-firstClickTime >=2000){
            Toast.makeText(this,"再按一次退出程序", Toast.LENGTH_SHORT).show();
            firstClickTime=System.currentTimeMillis();
        } else
            this.finish();
    }
    return false;
}
```

Android 数据存储解决方案篇

项目导引

在运行时,如果是第一次安装 CoffeeStore 项目,则会出现一个由几个广告图片切换的欢迎页面;如果不是第一次安装这个项目,则会直接进入首页,而不会运行这个启动欢迎页。那么如何判断用户是不是第一次安装这个 App 呢,安装项目的一些基本配置信息存放在何处呢?还有用户放到"收藏夹"中的店铺信息,即使当前手机没有网络,也能看到收藏夹里的商铺信息,收藏夹里收藏的店铺信息数据又是从哪里来的呢?还有,如果想播放手机 SD 卡里的 MP3 音乐,这些 MP3 音乐数据又存放在什么位置呢?

以上信息都是存储在 Android 系统的本地存储器里。本章将通过实现 CoffeeStore 的"启动页"以及"购物车"来讲解 Android 的本地存储技术。

运行时,商品信息等数据均来自服务器端程序的返回结果,因此,需要完成 Android 客户端程序和服务器端程序建立网络连接并进行网络通信,以及对返回的数据格式进行正确解析等操作。由于进行网络通信是耗时操作,如果将网络通信操作放在子线程中进行,由于 Android 限制仅能在主线程中更新 UI 界面,所以需要提供一个子线程和主线程进行通信的手段,除了前面介绍过的 Handler 之外,Android 中也提供了异步任务这种更轻量级、更简单的方式,完成子线程与主线程之间的消息传递。

另外,服务端程序返回的数据格式,也需要在 Android 客户端程序中进行正确的解析。目前较为常用的数据格式是 JSON 格式。

第 12 章将通过"用户登录"功能和"店铺列表"功能来讲解 Android 中的异步任务使

用，对服务端返回的 JSON 数据格式的解析以及利用 HttpURLConnection 进行网络通信的基本操作。

本篇将通过实现 CoffeeStore 的数据存储和传输来讲解 Android 的数据存储解决方案。Android 数据存储解决方案包括：

- 本地存储技术
- 网络存储技术

本地存储技术

本章概述

本章讲解 Android 的本地存储技术,包括简单数据存储类、Android 文件以及 SQLite 数据库的应用。

学习重点与难点

重点:

(1) SharedPreferences 存储类。
(2) 文件流操作 openFileOutput 和 openFileIntput。
(3) 读写 SD 卡中的数据。
(4) SQLiteOpenHelper 类。
(5) 数据库的增删改查操作。

难点:

(1) 读写 SD 卡中的数据。
(2) 数据库的查询操作。

学习建议

在理解 Android 的 3 种本地存储技术特点的基础上,要通过实践掌握每种存储技术用到的类库。数据库查询一直是初学的难点,要通过反复练习才能掌握。

11.1 简单数据存储类 SharedPreferences

11.1.1 SharedPreferences 的使用场合

很多开发软件需要提供软件参数设置功能,如常用的 QQ,用户可以设置是否允许陌生人添加自己为好友。对于保存软件配置参数,如果是 Windows 软件,通常会采用 ini 文件保存;如果是 J2SE 应用,会采用 properties 属性文件或 XML 保存。Android 平台提供了一个 SharedPreferences 类,是一个轻量级存储类,特别适合保存软件的配置参数。

使用 SharedPreferences 保存数据,背后是 XML 文件保存数据,文件存放在/data/data/<package name>/shared_prefs 目录下。

11.1.2 使用 SharedPreferences 存取数据

SharedPreferences 不仅能够保存数据,还能够实现不同应用程序间的数据共享。SharedPreferences 支持以下 3 种访问模式。

(1) 私有(MODE_PRIVATE=0):仅创建 SharedPreferences 的程序有权限对其进行读取或写入。

(2) 全局读(MODE_WORLD_READABLE=1):不仅创建程序可以对其进行读取或写入,其他应用程序也具有读取操作的权限,但没有写入操作的权限。

(3) 全局写(MODE_WORLD_WRITEABLE=2):所有程序都可以对其进行写入操作,但没有读取操作的权限。

使用 SharedPreferences 前,先定义 SharedPreferences 的访问模式,以下代码将访问模式定义为私有模式。

```
public static int MODE=MODE_PRIVATE;
```

有时需要将 SharedPreferences 的访问模式设定为既可以全局读,也可以全局写,这就需要将两种模式写成以下方式。

```
public static int MODE=Context.MODE_WORLD_READABLE+Context.MODE_WORLD_WRITEABLE;
```

除了定义 SharedPreferences 的访问模式,还要定义 SharedPreferences 的名称,这个名称也是 SharedPreferences 在 Android 文件系统中保存的文件名称。一般将 SharedPreferences 名称声明为字符串常量,这样可以在代码中多次使用。定义的代码如下。

```
public static final String PREFERENCE_NAME="SaveSetting";
```

使用 SharedPreferences 时,需要将访问模式和 SharedPreferences 名称作为参数传递到 getSharedPreferences() 函数,则可获取 SharedPreferences 实例。代码如下。

```
SharedPreferences sharedPreferences=getSharedPreferences(PREFERENCE_NAME, MODE);
```

获取到 SharedPreferences 实例后,可以通过 SharedPreferences.Editor 类修改 SharedPreferences,最后调用 commit() 函数保存修改内容。SharedPreferences 广泛支持各种基本数据类型,包括整型、布尔型、浮点型和长型等。SharedPreferences 对象读写键值对数据。通过 SharedPreferences 对象的键 key,获取到对应 key 的键值。不同类型的键值有不同的函数:getString、getBoolean、getInt、getFloat、getLong。以下示例代码表示通过 SharedPreferences 把一个人的基本信息保存到上面创建的 SaveSetting 文件中。

```
SharedPreferences.Editor editor=sharedPreferences.edit();
editor.putString("Name", "Tom");
editor.putInt("Age", 20);
```

```
editor.putFloat("Height", 1.81f);
editor.commit();
```

如果需要从已经保存的 SharedPreferences 中读取数据,同样是调用 getSharedPreferences()函数,并在函数的第 1 个参数中指明需要访问的 SharedPreferences 名称,最后通过 get<Type>()函数获取保存在 SharedPreferences 中的 NVP。get<Type>()函数的第 1 个参数是 NVP 的名称,第 2 个参数是在无法获取数值时使用的缺省值。下面的示例代码表示通过 SharedPreferences 读取保存在 SaveSetting 文件中的内容。

```
SharedPreferences sharedPreferences = getSharedPreferences(PREFERENCE_NAME, MODE);
String name=sharedPreferences.getString("Name","Default Name");
int age=sharedPreferences.getInt("Age", 20);
float height=sharedPreferences.getFloat("Height",1.81f);
```

11.2 Android 文件

从上一节可以知道,SharedPreferences 只能保存 key-value 对,而且只能读写字符串类型的数据。如果要读写更复杂的流数据(图像、音频、视频、压缩文件等),就需要使用本节介绍的文件流操作。Android 使用 Linux 的文件系统,开发人员可以建立和访问程序自身建立的私有文件,也可以访问保存在资源目录中的原始文件和 XML 文件,还可以将文件保存在 SD 卡等外部存储设备中。

11.2.1 文件数据的存储与读取

Android 系统允许应用程序创建仅能够自身访问的私有文件,文件保存在设备的内部存储器上,即 Android 系统下的/data/data/<package name>/files 目录中。Android 系统不仅支持标准 Java 的 IO 类和方法,还提供了能够简化读写流式文件过程的函数,这里主要介绍两个方法:openFileOutput()和 openFileInput()。

openFileOutput()函数为写入数据作准备而打开文件,如果指定的文件存在,直接打开文件,准备写入数据。如果指定的文件不存在,则创建一个新的文件,openFileOutput()方法的语法格式如下。

```
public FileOutputStream openFileOutput(String name, int mode)
```

第 1 个参数是文件名称,不能包含路径分隔符"/",第 2 个参数是操作模式,Android 系统支持 4 种文件操作模式,如表 11.1 所示。方法的返回值是 FileOutputStream 类型。

表 11.1　4 种文件操作模式

模　　式	说　　明
Context.MODE_PRIVATE=0	私有模式，默认操作模式，只能被应用本身访问，在该模式下，写入的内容会覆盖源文件内容
Context.MODE_APPEND=32768	追加模式，如果文件已经存在，则在文件的结尾处添加新数据
MODE_WORLD_READABLE=1	全局读模式，允许任何程序读取私有文件
MODE_WORLD_WRITEABLE=2	全局写模式，允许任何程序写入私有文件

Android 有一套自己的安全模式，当应用程序（.apk）安装时，系统会分配给它一个 userid，当该应用要访问其他资源（比如文件）时，就跟 userid 匹配。默认情况下，任何应用创建的文件都应该是私有的，其他程序无法访问。除非在创建时指定 MODE_WORLD_READABLE 或 MODE_WORLD_WRITEABLE，其他程序才能正确访问。

【例 11-1】　使用 FileOutputStream 写入二进制数据。

```
public void writeBinaryData(){

//写入二进制数据
String dataBinaryFileName="myBinaryData.data";
try{
        OutputStream stream=this.openFileOutput(dataBinaryFileName,Context.
        MODE_PRIVATE);
BufferedOutputStream bos=new BufferedOutputStream(stream);
byte[] data={1,2,3};
bos.write(data);
bos.close();
stream.close();
} catch(Exception e){
e.printStackTrace();
}
}
```

openFileInput()函数为读取数据作准备而打开文件，openFileInput()函数的语法格式如下。

```
public FileInputStream openFileInput(String name)
```

第 1 个参数也是文件名称，同样不允许包含路径分隔符的"/"，使用 openFileInput()函数打开已有文件。

【例 11-2】　使用 FileInputStream 读出二进制数据。

```
public  void readBinaryData(){
byte[] data=new byte[3];
String dataBinaryFileName="myBinaryData.data";
    try{
```

```
        InputStream inputStream=this.openFileInput(dataBinaryFileName);
BufferedInputStream bis=new BufferedInputStream(inputStream);
bis.read(data);
bis.close();
inputStream.close();

}catch(Exception e){
        Log.e("Tag",e.toString());
}
}
```

【例 11-3】 内部文件存储与读取实例。

布局文件 main.xml 的内容如下。

```xml
<?xml version="1.0" encoding="utf-8"?>
<LinearLayout xmlns:android="http://schemas.android.com/apk/res/android"
android:orientation="vertical"
android:layout_width="fill_parent"
android:layout_height="fill_parent"
>
<TextView  android:id="@+id/label"
android:layout_width="fill_parent"
android:layout_height="wrap_content"
android:text="@string/hello"
android:textSize="24sp"
/>
<EditText android:id="@+id/entry"

android:text="输入文件内容"
android:textSize="24sp"
android:layout_width="fill_parent"
android:layout_height="wrap_content">
</EditText>
<LinearLayout android:id="@+id/LinearLayout01"
android:layout_width="wrap_content"
android:layout_height="wrap_content">
<Button android:id="@+id/write"
android:text="写入文件"
android:textSize="24sp"
android:layout_width="wrap_content"
android:layout_height="wrap_content">
</Button>
<Button android:id="@+id/read"
android:text="读取文件"
android:textSize="24sp"
```

```xml
            android:layout_width="wrap_content"
            android:layout_height="wrap_content">
        </Button>
    </LinearLayout>
    <CheckBox android:id="@+id/append"
        android:text="追加模式"
        android:textSize="24sp"
        android:layout_width="wrap_content"
        android:layout_height="wrap_content">
    </CheckBox>
    <TextView android:id="@+id/display"
        android:text="文件内容显示区域"
        android:textSize="24sp"
        android:layout_width="fill_parent"
        android:layout_height="fill_parent"
        android:background="#FFFFFF"
        android:textColor="#000000" >
    </TextView>
</LinearLayout>
```

后台 InternalFileDemo.java 文件对应的内容如下。

```java
package neusoft.soft.storage;
import java.io.FileInputStream;
import java.io.FileNotFoundException;
import java.io.FileOutputStream;
import java.io.IOException;
import android.app.Activity;
import android.content.Context;
import android.os.Bundle;
import android.view.View;
import android.view.View.OnClickListener;
import android.widget.Button;
import android.widget.CheckBox;
import android.widget.EditText;
import android.widget.TextView;
import com.example.fuli.coffeestorebak.R;
public class InternalFileDemo extends Activity {
    private final String FILE_NAME="fileDemo.txt";
    private TextView labelView;
    private TextView displayView;
    private CheckBox appendBox ;
    private EditText entryText;
    @Override
    public void onCreate(Bundle savedInstanceState){
```

```java
        super.onCreate(savedInstanceState);
        setContentView(R.layout.main);
        labelView=(TextView)findViewById(R.id.label);
        displayView=(TextView)findViewById(R.id.display);
        appendBox=(CheckBox)findViewById(R.id.append);
         entryText=(EditText)findViewById(R.id.entry);
        Button writeButton=(Button)findViewById(R.id.write);
        Button readButton=(Button)findViewById(R.id.read);
        writeButton.setOnClickListener(writeButtonListener);
        readButton.setOnClickListener(readButtonListener);
        entryText.selectAll();
        entryText.findFocus();
    }
    OnClickListener writeButtonListener=new OnClickListener(){
      @Override
      public void onClick(View v){
         FileOutputStream fos=null;
          try {
              if(appendBox.isChecked()){
                  fos=openFileOutput(FILE_NAME,Context.MODE_APPEND);
              }
              else {
                  fos=openFileOutput(FILE_NAME,Context.MODE_PRIVATE);
              }

              String text=entryText.getText().toString();
              fos.write(text.getBytes());
              labelView.setText("文件写入成功,文件长度"+text.length());
              entryText.setText("");
        } catch(FileNotFoundException e){
             e.printStackTrace();
        }
        catch(IOException e){
            e.printStackTrace();
        }
        finally{
            if(fos !=null){
              try {
                 fos.flush();
                 fos.close();
              } catch(IOException e){
                 e.printStackTrace();
              }
            }
```

```java
            }
        }
    };

    OnClickListener readButtonListener=new OnClickListener(){
        @Override
        public void onClick(View v){
            displayView.setText("");
            FileInputStream fis=null;
            try{
                fis=openFileInput(FILE_NAME);
                if(fis.available()==0){
                    return;
                }
                byte[] readBytes=new byte[fis.available()];
                while(fis.read(readBytes)!=-1){
                }
                String text=new String(readBytes);
                displayView.setText(text);
                labelView.setText("文件读取写入成功,文件长度"+text.length());
            } catch(FileNotFoundException e){
                e.printStackTrace();
            }
            catch(IOException e){
                e.printStackTrace();
            }

        }
    };
}
```

程序运行结果如图 11.1 所示。在文本框中输入要写入文件的内容,单击"写入文件"按钮,则会把文本框内容写入文件 fileDeno.txt 中。若选中复选框"追加模式",则新写入的内容会追加到原有文件末尾,否则会删除源文件后把新内容写入文件中。

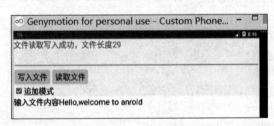

图 11.1 读写文件实例运行效果

11.2.2 读写 SD 卡中的文件

使用 openFileOutput 方法保存文件,文件是存放在手机空间里的。一般手机的存储空间不是很大,存一些小文件还可以,如果要存放像视频这样的大文件,需要使用 SD 卡。

Android 模拟器支持 SD 卡模拟,建立模拟器时可以选择 SD 卡的容量。正确加载 SD 卡后,SD 卡中的目录和文件被映射到/mnt/sdcard 目录下。因为用户可以加载或卸载 SD 卡,所以在编程访问 SD 卡前,首先需要检测/mnt/sdcard 目录是否可用,如果不可用,说明设备中的 SD 卡已经被卸载。如果可用,则直接通过使用标准的 java.io.File 类进行访问。获取 SD 卡基本信息的示例代码如下。

```
//判断是否有插入存储卡
if(Environment.getExternalStorageState().equals(Environment.MEDIA_MOUNTED)){
//获取 SD 卡所在的目录
File path=Environment.getExternalStorageDirectory();
//获得 sdcard 文件路径
   StatFs statFs=new StatFs(path.getPath());
```

在程序中访问 SD 卡,需要申请 SD 卡的权限。在 AndroidMenifest.xml 中加入访问 SD 卡的权限,代码如下。

```
<uses-permission android:name="android.permission.WRITE_EXTERNAL_STORAGE"/>
<uses-permission android:name="android.permission.MOUNT_UNMOUNT_FILESYSTEMS"
/>
```

【例 11-4】 遍历 SD 卡中的所有文件及文件目录。

Listview 的布局文件 folder_list.xml 内容如下。

```
<?xml version="1.0" encoding="utf-8"?>
<LinearLayout xmlns:android="http://schemas.android.com/apk/res/android"
android:id="@+id/linear_layout"
android:layout_width="fill_parent"
android:layout_height="60dp"
android:orientation="horizontal">
<ImageView
android:id="@+id/image_view"
android:layout_width="60dp"
android:layout_height="60dp"
android:layout_gravity="center_vertical">
</ImageView>
<TextView
android:id="@+id/folder_name"
android:layout_width="fill_parent"
android:layout_height="20dp"
android:layout_gravity="center_vertical"
android:textSize="16sp" />
```

```xml
</LinearLayout>
```

主界面的布局文件 main.xml 内容如下。

```xml
<?xml version="1.0" encoding="utf-8"?>
<LinearLayout xmlns:android="http://schemas.android.com/apk/res/android"
android:layout_width="fill_parent"
android:layout_height="fill_parent"
android:orientation="vertical">

<LinearLayout
android:layout_width="fill_parent"
android:layout_height="wrap_content"
android:orientation="horizontal">

<TextView
android:id="@+id/text_view"
android:layout_width="fill_parent"
android:layout_height="wrap_content"
android:layout_gravity="center_vertical"
android:textSize="18px"></TextView>
</LinearLayout>

<ListView
android:id="@+id/listView"
android:layout_width="fill_parent"
android:layout_height="fill_parent"></ListView>
</LinearLayout>
```

TraverseFolder.java 文件的内容如下。

```java
package neusoft.soft.storage;
import android.app.Activity;
import android.app.AlertDialog;
import android.content.DialogInterface;
import android.content.DialogInterface.OnClickListener;
import android.content.Intent;
import android.os.Bundle;
import android.os.Environment;
import android.view.View;
import android.widget.AdapterView;
import android.widget.AdapterView.OnItemClickListener;
import android.widget.ListView;
import android.widget.SimpleAdapter;
import android.widget.TextView;
import com.example.fuli.coffeestorebak.R;
```

```java
import java.io.File;
import java.util.ArrayList;
import java.util.HashMap;
import java.util.List;
public class TraverseFolder extends Activity {
    private TextView textView=null;              //用于显示目录结构的TextView组件对象
    private File[] files=null;                   //File数组
    private ListView listView=null;              //用于显示文件的ListView组件对象

    @Override
    public void onCreate(Bundle savedInstanceState){
        super.onCreate(savedInstanceState);
        setContentView(R.layout.main);
        //实例化ListView组件对象
        listView=(ListView)findViewById(R.id.listView);
        //实例化TextView组件对象
        textView=(TextView)findViewById(R.id.text_view);
        //调用获取手机SD卡的存储状态
        boolean sdStatus=getStorageState();
            if(!sdStatus){                       //判断SD卡的存储状态,如果是false,提示并结束本程序
        AlertDialog alertDialog=new AlertDialog.Builder(
                TraverseFolder.this).create();   //创建AlertDialog对象
        alertDialog.setTitle("提示信息");         //设置信息标题
        alertDialog.setMessage("未安装SD卡,请检查你的设备");  //设置信息内容
                //设置确定按钮,并添加按钮监听事件
        alertDialog.setButton("确定", new OnClickListener(){

        @Override
        public void onClick(DialogInterface arg0, int arg1){
                TraverseFolder.this.finish();    //结束应用程序
        }
            });
        alertDialog.show();                      //设置弹出提示框
        }
            else{
            Intent intent=getIntent();           //获取Intent
        CharSequence charSequence=intent.getCharSequenceExtra("filePath");
                                                 //获取CharSequence对象
        if(charSequence !=null){                 //判断CharSequence对象是否为空,
                                                 //为空就获取SD卡根目录,否则就获
                                                 //取传过来的文件目录
        File file=new File(charSequence.toString());  //实例化File
        textView.setText(file.getPath());        //更新TextView组件显示的目录结构
        files=file.listFiles();                  //获取该目录的所有文件及目录
```

```java
    } else {
            File sdCardFile=Environment.getExternalStorageDirectory();
                                            //获取 SD 卡根目录 File 对象
textView.setText(sdCardFile.getPath());     //设置 TextView 组件显示的目录结构
files=sdCardFile.listFiles();               //获取 SD 卡根目录的所有文件及目录
    }

        List<HashMap<String, Object>>list=getList(files);
                                            //调用获取相应的集合
setAdapter(list, files);                    //调用构造适配器并为 ListView 添加适配器

listView.setOnItemClickListener(new OnItemClickListener(){
                                            //为 ListView 添加单击监听

@Override
public void onItemClick(AdapterView<?>arg0, View arg1, int arg2,
                    long arg3){
if(files[arg2].isDirectory()){              //判断所单击的文件是否文件夹
File[] childFiles=files[arg2].listFiles();  //获取该单击文件夹下的所有文件及文件夹
if(childFiles != null && childFiles.length >=0){ //判断该单击文件夹数组不为空
Intent intent=new Intent();                 //初始化 Intent
intent.setClass(TraverseFolder.this,
TraverseFolder.class);                      //指定 intent 对象启动的类
intent.putExtra("filePath", files[arg2].getPath());    //函数传递
startActivity(intent);                      //启动新的 Activity
}
        }
    }
});
    }
}
    //构造适配器并为 ListView 添加适配器
public void setAdapter(List<HashMap<String, Object>>list, File[] files){
     SimpleAdapter simpleAdapter=new SimpleAdapter(TraverseFolder.this,
list, R.layout.folder_list, new String[]{"image_view",
"folder_name"}, new int[]{R.id.image_view,
R.id.folder_name});                         //实例化 SimpleAdapter

listView.setAdapter(simpleAdapter);         //为 ListView 添加适配器
this.files=files;                           //给当前 File 数组赋值
    }

    //获取手机 SD 卡的存储状态
public boolean getStorageState(){
```

```
if(Environment.getExternalStorageState().equals(
            Environment.MEDIA_MOUNTED)){        //判断手机 SD 卡的存储状态
return true;
} else {
return false;
}
    }
public List<HashMap<String, Object>>getList(File[] files){
      List<HashMap<String, Object>>list=new ArrayList<HashMap<String, Object
>>();                                          //创建 List 集合
for(int i=0; i<files.length; i++){              //循环 File 数组
HashMap<String, Object>hashMap=new HashMap<String, Object>();
                                                //创建 HashMap
if(files[i].isDirectory()){                     //判断该文件是否文件夹
hashMap.put("image_view", R.drawable.dir1);     //往 HashMap 中添加文件夹图片
} else {
        hashMap.put("image_view", R.drawable.file2);
                                                //往 HashMap 中添加文件图片
}
        hashMap.put("folder_name", files[i].getName());
                                                //往 HashMap 中添加文件名
list.add(hashMap);                              //将 HashMap 添加到 List 集合
}
return list;                                    //返回 List 集合
}
}
```

11.2.3 读写资源文件

开发人员除了可以在内部和外部存储设备上读写文件以外,还可以访问在/res/raw 和/res/xml 目录中的原始格式文件和 XML 文件,这些文件是程序开发阶段在工程中保存的文件。原始格式文件可以是任何格式的文件,例如视频格式文件、音频格式文件、图像文件或数据文件等。在应用程序编译和打包时,/res/raw 目录下的所有文件都会保留原有格式不变。而/res/xml 目录下一般用来保存格式化数据的 XML 文件,则会在编译和打包时将 XML 文件转换为二进制格式,以降低存储器占用空间,提高访问效率,运行应用程序时会以特殊的方式访问。

存放在 assets 及 res 下的文件,可以使用 Activity 中的方法获取 InputStream 对象。

- context.getClass().getClassLoader().getResourceAsStream("assets/"+资源名); 返回某个文件名对应的 assets 目录下文件的访问流。
- getResources.openRawResource(int)   返回 ResourceId 对应的 res 目录文件的访问流。

【例 11-5】 XML 资源文件的读取与解析。

在 res 目录下创建 xml 目录,并在 xml 目录下创建一个 person.xml 文件,内容如下。

```xml
<?xml version="1.0" encoding="UTF-8"?>
<persons>
<person id="1">
<name>zhansan</name>
<age>23</age>
</person>
<person id="2">
<name>li</name>
<age>40</age>
</person>
</persons>
```

Person 类的定义如下。

```java
public class Person {
private String id;
    private String name;
    private int age;
    public String getId(){
return id;
}
public void setId(String id){
this.id=id;
}
public String getName(){
return name;
}
public void setName(String name){
this.name=name;
}
public int getAge(){
return age;
}
public void setAge(int age){
this.age=age;
}

}
```

使用 XmlPullParse 类解析 XML 文件的代码如下。

```java
List<Person>persons=null;
Person p=null;
//解析 XML
```

```java
//1.获取解析器
XmlPullParser parser=getResources().getXml(R.xml.person);
//2.获取事件类型
try {
int type=parser.getEventType();                //获取时间类型
while(type !=  XmlPullParser.END_DOCUMENT){    //如果未到文档末尾,开始解析
switch(type){
case XmlPullParser.START_DOCUMENT://如果文档开始标记,创建保存Person集合的List对象
persons=new ArrayList<Person>();
            break;
        case XmlPullParser.START_TAG:
            String tagName=parser.getName();
            if(tagName.equals("person")){
                p=new Person();
p.setId(parser.getAttributeValue(0));
}
if(tagName.equals("name")){
            p.setName(parser.nextText());
}
if(tagName.equals("age")){
            p.setAge(Integer.parseInt(parser.nextText()));
}
break;
        case XmlPullParser.END_TAG:
            tagName=parser.getName();
            if(tagName.equals("person")){
                persons.add(p);
p=null;
}
break;
}
//驱动指向下一个节点(元素和文本节点)
type=parser.next();
}

for(int i=0; i<persons.size(); i++){
        Person p1=persons.get(i);
System.out.println(p1.getId()+":" +p1.getName()+":"+p1.getAge());
}
} catch(XmlPullParserException e){
//TODO Auto-generated catch block
e.printStackTrace();
} catch(IOException e){
```

```
//TODO Auto-generated catch block
e.printStackTrace();
}
```

程序分析：用 getName() 函数获得元素的名称，用 getAttributeCount() 函数获取元素的属性数量，通过 getAttributeName() 函数得到属性名称。XML 事件类型如下。

- START_TAG 读取到标签开始标志。
- TEXT 读取文本内容。
- END_TAG 读取到标签结束标志。
- END_DOCUMENT 文档末尾。

11.3 SQLite 数据库

11.3.1 SQLite 数据库存储数据概述

Android 平台集成了一个嵌入式轻量级关系型数据库——SQLite。SQLite 3 支持 NULL、INTEGER、REAL（浮点数字）、TEXT（字符串文本）和 BLOB（二进制对象）数据类型。虽然支持的类型只有 5 种，但 SQLite 3 也接受 varchar(n)、char(n)、decimal(p,s) 等数据类型，只不过运算或保存时会自动转成对应的 5 种数据类型。SQLite 最大的特点是可以保存任何类型的数据到任何字段中，无论这列声明的数据类型是什么。例如，可以在 Integer 类型的字段中存放字符串，或者在字符型字段中存放日期型值。但有一种情况例外：定义为 INTEGER PRIMARY KEY 的字段只能存储 64 位整数，当向这种字段中保存除整数外的数据时，会产生错误。SQLite 可以解析大部分标准 SQL 语句。

SQLite 数据库相关类/接口如下。

(1) SQLiteOpenHelper 类：是一个辅助抽象类，主要用来进行数据库的创建和版本管理，通常需要创建子类继承它。

(2) SQLiteDatabase 类：一个 SQLiteDatabase 的实例代表一个 SQLite 数据库，通过 SQLiteDatabase 实例的方法可以执行 SQL 语句，从而实现对数据库的增、删、改、查等操作。

(3) Cursor 接口：Cursor 是 Android 非常有用的接口，通过 Cursor 可以对数据库的查询结果集进行随机的读写访问。

(4) ContentValues 类：ContentValues 存储一些名值对。提供数据库的列名、数据映射信息，ContentValues 对象代表了数据库的一行数据。

11.3.2 使用 SQLiteOpenHelper 类管理数据库版本

android.database.sqlite.SQLiteDatabase 是 Android SDK 中操作数据库的核心类之一，使用 SQLiteDatabase 可以打开数据库，也可以对数据库进行操作。然而，为了数据库升级的需要以及使用方便，往往使用 SQLiteOpenHelper 的子类来完成创建、打开数据库及各种数据库的操作。

SQLiteOpenHelper 是一个抽象类，该类中有如下两个方法，SQLiteOpenHelper 的子类必须实现这两个方法。

```
public void onCreate(SQLiteDatabase db);
public void onUpgrade(SQLiteDatabase db,int oldVersion,int new Version);
```

调用 SQLiteOpenHelper 的 getWritableDatabase 或 getReadableDatabase 方法获取用于操作数据库的 SQLiteDatabase 实例时，SQLiteOpenHelper 会自动检测数据文件是否存在。如果数据库文件存在，会打开这个数据库，这种情况下并不会调用 onCreate 方法，如果数据库文件不存在，SQLiteOpenHelper 首先会创建一个数据库文件，然后打开这个数据库，最后调用 onCreate 方法。因此，onCreate 方法一般用来在新创建的数据库中简历表、视图等数据库组件。也就是说，onCreate 方法在数据库文件第一次被创建时调用。onUpgrade 方法在数据库的版本发生变化时会调用，一般在软件升级时才需要改变版本号，而数据库的版本是由程序员控制的。

【例 11-6】 SQLiteOpenHelper 使用实例。本例定义了一个 DBService 类，该类是 SQLiteOpenHelper 的子类，用于创建数据库、建立 t_test 表和查询 t_test 表中的记录。

```java
import android.content.Context;
import android.database.Cursor;
import android.database.sqlite.SQLiteDatabase;
import android.database.sqlite.SQLiteOpenHelper;
import java.util.Random;
public class DBService extends SQLiteOpenHelper
{
private final static int DATABASE_VERSION=1;
    private final static String DATABASE_NAME="test.db";

@Override
public void onCreate(SQLiteDatabase db)
   {

      String sql="CREATE TABLE [t_test]("+"[_id] AUTOINC,"
+"[name] VARCHAR(20)NOT NULL ON CONFLICT FAIL,"
+"CONSTRAINT [sqlite_autoindex_t_test_1] PRIMARY KEY([_id]))";

db.execSQL(sql);
Random random=new Random();
     for(int i=0; i<20; i++)
     {
        String s="";
        for(int j=0; j<10; j++)
        {
char c=(char)(97+random.nextInt(26));
s +=c;
```

```
            db.execSQL("insert into t_test(name)values(?)", new Object[]
            { s });
        }

    }

    public DBService(Context context)
    {
    super(context, DATABASE_NAME, null, DATABASE_VERSION);
    }

    @Override
    public void onUpgrade(SQLiteDatabase db, int oldVersion, int newVersion)
        {
        }

    public Cursor query(String sql, String[] args)
        {                    //调用getReadableDatabase方法获取SQLiteDatabase实例

            SQLiteDatabase db=this.getReadableDatabase();
    Cursor cursor=db.rawQuery(sql, args);
            return cursor;
    }
    }
```

程序分析：DBService 类创建了一个 test.db 数据库文件，并在该文件中创建了 t_test 表，该表包含了两个字段：_id 和 name，其中_id 是自增字段，并且是主索引。

11.3.3 使用 SQLiteDatabase 操作数据库

Android 提供了一个名为 SQLiteDatabase 的类，该类封装了一些操作数据库的 API，使用该类可以对数据进行添加、查询、更新和删除操作。execSQL 方法可以执行 Insert、Delete、Update 和 CREATE TABLE 之类有更改行为的 SQL 语句；rawQuery 方法可以执行 Select 语句。

【例 11-7】 支持使用占位符参数(?)的 execSQL 方法。

```
SQLiteDatabase db=...;
db.execSQL("insert into person(name,age)values(?,?)", new Object[]{"张三", 18});
db.close();
```

SQLiteDatabase 类常用方法还有以下几个。
(1) 创建或打开数据库的静态方法。

openDatabase(String path, SQLiteDatabase.CursorFactory factory,int flags)

功能：打开指定路径的数据库文件。

参数 path：指定路径的数据库文件。

参数 factory：用于构造查询时返回的 Cursor 对象。

参数 flags：打开模式,包括 OPEN_READONLY(只读方式)、OPEN_READWRITE(可读可写)和 CREATE_IF_NECESSARY(当数据库文件不存在时,创建该数据库)。

openOrCreateDatabase(String path，SQLiteDatabase. CursorFactory factory)相当于用 openDatabase()方法打开模式为 CREATE_IF_NECESSARY 的情形。

(2) update()：修改表中数据。有以下 4 个参数。

第 1 个参数：String,数据表名称。

第 2 个参数：ContentValues,ContentValues 对象。

第 3 个参数：String,where 子句,相当于 SQL 语句 where 后面的语句,? 号是占位符。

第 4 个参数：String[],占位符的值。

update()函数的返回值表示数据库表中被更新的数据数量。

例：sqliteDatabase. update("user"，values，"id＝?"，new String[]｛ "1" ｝);

(3) delete()：删除表中数据。有以下 3 个参数。

第 1 个参数：String,数据表名称。

第 2 个参数：String,条件语句。

第 3 个参数：String[],条件值。

delete()函数的返回值表示被删除的数据数量。

例：sqliteDatabase. delete("user"，"id＝?"，new String[]{"1"});

(4) insert()：插入数据。有以下 3 个参数。

第 1 个参数：String,数据表名称。

第 2 个参数：SQL 不允许一个空列,如果 ContentValues 是空的,那么这一列被明确指明为该参数值。

第 3 个参数：ContentValues 对象,为插入值。

insert()函数的返回值是新数据插入的位置,即 ID 值。

例：sqliteDatabase. insert("user"，null，valueinsert);

(5) query()：查询数据。有以下 7 个参数。

第 1 个参数：String,数据表名称。

第 2 个参数：String[],返回的属性列名称。

第 3 个参数：String,查询条件。

第 4 个参数：String g[],如果在查询条件中使用通配符(?),则需要在这里定义替换符的具体内容。

第 5 个参数：String,分组方式。

第 6 个参数：String,定义组的过滤器。

第 7 个参数：String,排序方式。

数据库查询结果的返回值并不是数据集合的完整复制,而是返回数据集的指针,这个指针就是 Cursor 类。Cursor 类支持在查询结果的数据集合中以多种方式移动，并能够获取数据集合的属性名称和序号，具体的方法和说明可以参考表 11.2。

表 11.2 Cursor 类的公有方法

函　　数	说　　明
moveToFirst	将指针移动到第一条数据上
moveToNext	将指针移动到下一条数据上
moveToPrevious	将指针移动到上一条数据上
getCount	获取集合的数据数量
getColumnIndexOrThrow	返回指定属性名称的序号，如果属性不存在则产生异常
getColumnName	返回指定序号的属性名称
getColumnNames	返回属性名称的字符串数组
getColumnIndex	根据名称返回序号
moveToPosition	将指针移动到指定的数据上
getPosition	返回当前指针的位置

【例 11-8】将【例 11-6】的 DBService 类创建的数据库表中的数据显示到 ListView 上。将表中数据显示在 ListView、GridView 等控件中，虽然可以直接使用 BaseAdapter 处理，但工作量较大。Android 提供了一个专用于数据绑定的 Adapter 类：SimpleCursorAdapter 类。它的用法与 SimpleAdapter 非常相似，只是将数据源从 List 对象转换成了 Cursor 对象。

```
importandroid.os.Bundle;
import android.app.ListActivity;
import android.database.Cursor;
import android.widget.SimpleCursorAdapter;
public class Main extends ListActivity
{
public void onCreate(Bundle savedInstanceState)
    {
super.onCreate(savedInstanceState);
DBService dbService=new DBService(this);
Cursor cursor=dbService.query("select * from t_test",null);
SimpleCursorAdapter simpleCursorAdapter=new SimpleCursorAdapter(this,
android.R.layout.simple_expandable_list_item_1, cursor,
        new String[]
        {"name" }, new int[]
        { android.R.id.text1},1);

setListAdapter(simpleCursorAdapter);
```

}
　　}

程序分析：本程序的 Main 类是 ListActivity 的子类。在 Main.onCreate 方法中创建了 DBService 对象，然后通过 query 方法查询出 t_test 表中的所有记录，并返回 Cursor 对象。最后将这个 Cursor 对象传入 SimpleCursorAdapter 类的构造方法，并将 SimpleCursorAdapter 对象与控件 ListView 绑定。绑定数据时，Cursor 对象返回的记录集中必须包含一个名为 _id 的字段，否则无法完成数据绑定。SimpleCursorAdapter 类构造方法的第 4 个参数表示返回的 Cursor 对象中的字段名，第 5 个参数表示要显示该字段的控件 ID，该控件在第 2 个参数指定的布局文件中定义。运行本例后，将显示如图 11.2 所示的效果。

图 11.2　使用 SimpleCursorAdapter 操作数据

　　例 11-6 的例子建立了 SQLite 数据库，数据库文件被放到哪个目录了呢？如果使用 SQliteOpenHelper.getReadableDatabase 或 getWritableDatabase 方法获得 SQLiteDatabase 对象，系统会在手机内存的 /data/data/＜package name＞/databases 目录中创建或寻找数据库。

11.3.4　一起发布数据库与应用程序

　　在前面的例子中，都是程序第一次启动时创建数据库，也就是说，数据库文件是由应用程序创建的。一般初始状态的数据库中没有记录，就算有记录，也是由应用程序创建数据库时添加的。应用程序发布时并不包含带数据的数据库，但在很多情况下，应用程序需要连同带数据的数据库一起发布，这就需要通过某种机制打开 apk 文件中的数据库。

　　要满足上述需求，一般要解决如下两个技术问题。

　　(1) 如何将数据库文件连同应用一起发布。

　　(2) 如何打开与应用程序一起发布的数据库。

　　第一个问题好解决，可以事先利用一些数据库管理工具在 PC 上建立一个数据库文件，并手工或通过程序向数据表中添加相应的记录，然后将该数据库文件复制到＜Android Studio 工程目录＞/res/raw 目录或 assets 目录中。

　　第二个问题如何解决呢？发布 apk 时，数据库文件被打包在 apk 文件中，那么如何打开这个 apk 呢？实际上并不能直接打开 apk 包中的数据库，因此，第一次运行程序时，需要将数据库文件复制到内存或 SD 卡的相应目录。复制的方法也很简单，使用 Context 读取资源文件的方法获得 InputStream 对象。有了 InputStream 对象，复制文件就简单了。

11.4 项目实战：CoffeeStore 启动页安装信息的存取

11.4.1 项目分析

每次启动 CofffeeStore 项目，都要判断用户是否首次安装本 Apps，如果是第一次安装，需要经过 3 个切换广告，如果不是，就跳过广告，直接进入主页面。若想实现上述效果，需每次启动项目时读取项目的安装信息，如果不是第一次安装，则直接跳到首页，如果是第一次安装，则跳到欢迎页，同时需要把项目的安装信息保存下来。项目是否首次安装的配置信息可以使用 SharedPreference 存储。

11.4.2 项目实现

首先创建 3 个 Activity，分别表示项目启动页、欢迎页和首页，3 个 Activity 的名称如图 11.3 所示。

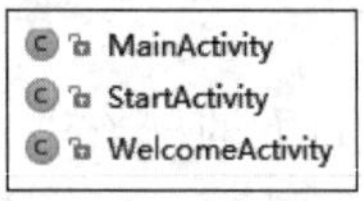

在 StartActivity 中使用 SharedPreference 保存和读取是否第一次安装的配置信息，代码如下。

图 11.3　Activity 组成

```
package neusoft.soft.coffeestore.view;
import android.app.Activity;
import android.content.Intent;
import android.content.SharedPreferences;
import android.content.SharedPreferences.Editor;
import android.os.Bundle;
public class StartActivity extends Activity {
@Override
protected void onCreate(Bundle savedInstanceState){
//TODO Auto-generated method stub
super.onCreate(savedInstanceState);
//该类主要用来判断用户是不是第一次安装本 App,如果是第一次安装,需要经过 3 个
viewfilper 切换广告,如果不是就跳过广告,直接进入主页面
SharedPreferences sharedata=getSharedPreferences("config",
        this.MODE_PRIVATE);
String data=sharedata.getString("time","");
    if(data.equals("secondTime")){
        Intent intent2=new Intent(this,MainActivity.class);
startActivity(intent2);
}else{
        SharedPreferences sp=this.getSharedPreferences("config",
            this.MODE_PRIVATE);
Editor editor=sp.edit();
editor.putString("time","secondTime");
editor.commit();
```

```
Intent intent1=new Intent(this,WelcomeActivity.class);
startActivity(intent1);
        }
    }
}
```

11.4.3　项目说明

StartActivity 类主要是用来判断用户是不是第一次安装本 App,如果是第一次安装,需要经过 3 个切换广告,如果不是就跳过广告,直接进入主页面。在 onCreate 方法中首先读取配置文件 config 中 time 的值,若为空表示第一次安装,保存 time 的值,并跳到欢迎页,否则表示已安装过,则跳到主页。

11.5　项目实战：读取数据库文件

11.5.1　项目分析

在本项目中,使用 SQLite 来存放"收藏夹"中收藏的店铺信息。为了能将 SQLite 数据库 coffeeshop.sqlite 文件与 apk 文件一起发布,可以把数据库文件 coffeeshop.sqlite 复制到 Android Studio 工程中的 assets 目录中。所有在 assets 目录中的文件不会被压缩,这样可以直接提取该目录中的文件,然后打开该目录中的数据库文件。

Android 中不能直接打开 assets 目录中的数据库文件,而需要在程序第一次启动时将该文件复制到手机内存或 SD 卡的某个目录中,然后再打开该数据库文件。复制的基本方法是使用 context.getClass().getClassLoader().getResourceAsStream 获得 assets 目录中资源的 InputStream 对象,然后将该 InputStream 对象中的数据写入其他目录中的相应文件中。在 Android SDK 中可以使用 SQLiteDatabase.openOrCreateDatabase 方法打开任意目录中的 SQLite 数据库文件。

店铺信息包括店铺名称、店铺图片、店铺地址、店铺电话,设计数据库表结构如表 11.3 所示。

表 11.3　店铺信息表结构

字 段 描 述	数 据 类 型	主　　键	可 否 为 空	描　　述
store_id(咖啡店编号)	int(11)	是	否	
store_name(咖啡店名称)	varchar(50)		否	
store_phone(咖啡店电话)	varchar(50)		否	
store_address(咖啡店地址)	varchar(50)		否	
store_img(咖啡店图片)	varchar(50)		否	

11.5.2　项目实现

Android Studio 工程里没有自带的 assets 目录,需要手动添加 assets 目录,assets 目

录必须放在 java 和 res 同级目录下，如图 11.4 所示。

coffeeshop.sqlite 数据库文件可以使用 SQLiteExpertPersSetup 工具创建，也可使用 NavicatLite 工具创建。使用 NavicatLite 工具创建的步骤如下。

(1) 打开 Navicat，选择文件→新建连接→SQLite，出现如图 11.5 所示的对话框。

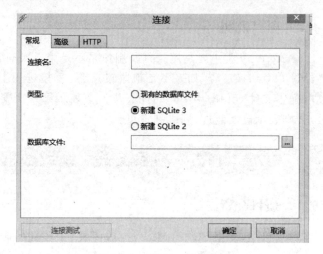

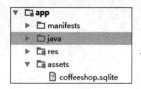

图 11.4　assets 目录位置　　　　　图 11.5　SQLite 连接界面

(2) 设置数据库文件的存储路径和文件名，选择"新建 SQLite 3"，单击"确定"按钮后即可创建数据库。

(3) 创建数据库表。

读取 coffeeshop.sqlite 数据库文件，并复制到内存数据库中，代码如下。

```
package neusoft.soft.storage;
import android.app.Activity;
import android.content.Context;
import android.content.pm.ApplicationInfo;
import android.content.pm.PackageManager;
import android.os.Bundle;
import java.io.File;
import java.io.FileOutputStream;
import java.io.IOException;
import java.io.InputStream;
public class CoffeeDBActivity extends Activity {
final String DB_NAME="coffeeshop";                    //数据库文件名
ApplicationInfo applicationInfo;
String databasePath;
    final String DB_DIR="databases";                  //数据库的存放路径
private void init(Context context){
        String packageName=context.getPackageName();//获取挡前应用程序的包名
try {
applicationInfo= context.getPackageManager().getApplicationInfo(packageName,
```

```java
                    PackageManager.GET_META_DATA);              //获取当前应用程序的元数据

    String dbDir=applicationInfo.dataDir+File.separator+DB_DIR;
                                                                //设置数据库的存放目录
    File file=new File(dbDir);
                if(!file.exists()){                             //若路径不存在,则创建路径

    file.mkdir();
    }
    databasePath=applicationInfo.dataDir+File.separator+DB_DIR+File.separator+
    DB_NAME;                                                    //存放数据库的完整路径

    } catch(PackageManager.NameNotFoundException e){
            }
        }

    private void initDB(){

    new Thread(new Runnable(){

    @Override
    public void run(){                                          //开启线程读取 assets 资源
    目录下的数据文件,并复制到内存的 data/data/packageName/databases 目录下
    try {
    //打开 assets 资源目录下的文件
    InputStream inputStream = CoffeeDBActivity.this.getClass().getClassLoader().
    getResourceAsStream("assets/"+"coffeeshop.sqlite");
    //若要复制的文件不存在,则创建一个文件
    if(databasePath!=null){
                    File file=new File(databasePath);

                    if(!file.exists()){
                        file.createNewFile();       //创建数据库文件
    }
                    FileOutputStream outputStream=new FileOutputStream(file);

                    byte[] buffer=new byte[1024 * 4];
                    int count=0;
    //读取资源目录下的数据库文件,复制到内存数据库文件中
    while((count=inputStream.read(buffer))!=-1){

                        outputStream.write(buffer, 0, count);
    }
                    outputStream.close();
```

```
                }
                        inputStream.close();
            } catch(IOException e){
                        e.printStackTrace();
            }
                }
        }).start();
}
@Override
protected void onCreate(Bundle savedInstanceState){
    super.onCreate(savedInstanceState);
    init(this);
    initDB();
}
```

11.5.3 项目说明

由于读取数据库操作是个耗时操作,所以本例开启线程来读取资源目录下的数据库文件,并把它复制到内存中。程序运行后,会把资源目录 assets 下的数据库文件复制到内存 data/data/目录下,如图 11.6 所示。

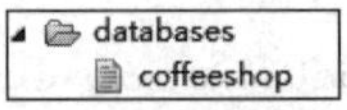

图 11.6 数据库在内存目录中的显示结构

11.6 项目实战：CoffeeStore 项目中本地收藏夹的实现

11.6.1 项目分析

在本项目中,购物车里的商品信息是一种表结构,因此采用数据库来存储。在 Android 中系统创建的数据库存放在/data/data/<package name>/sqlite 目录里。

11.6.2 项目实现

收藏夹 Activity 的代码如下。

```
import android.app.Activity;
import android.content.Context;
import android.content.Intent;
import android.content.pm.ApplicationInfo;
import android.content.pm.PackageManager;
import android.content.pm.PackageManager.NameNotFoundException;
import android.graphics.Color;
```

```java
import android.os.Bundle;
import android.view.View;
import android.view.ViewGroup;
import android.widget.AdapterView;
import android.widget.AdapterView.OnItemClickListener;
import android.widget.ListView;
import android.widget.SimpleAdapter;
import android.widget.TextView;
import java.io.File;
import java.util.ArrayList;
import java.util.HashMap;
import java.util.List;
import java.util.Map;
import cn.edu.neusoft.software.mobile.activitydemo.R;
public class FavoriteActivity extends Activity
{
    List<HashMap<String, String>>data;
    final String DB_DIR="databases";
    final String DB_NAME="coffeeshop";
    private ListView list;
    ApplicationInfo applicationInfo;
    String databasePath;
    DBUtil dbUtil;
    Shop[] shops;
    @Override
    protected void onCreate(Bundle savedInstanceState){
        super.onCreate(savedInstanceState);
        setContentView(R.layout.shop_layout);
        list=(ListView)findViewById(R.id.listshop);
        init(this);
        dbUtil=DBUtil.getInstance(databasePath);
        dbUtil.openDB();
        showAllShops();

    }

    @Override
    protected void onRestart(){
        //TODO Auto-generated method stub
        super.onRestart();
        showAllShops();
    }

    public void showAllShops(){
```

```java
            shops=dbUtil.queryAllShop();
            data=new ArrayList<HashMap<String, String>>();
            for(int i=0; i<shops.length; i++){
               HashMap<String, String>map=new HashMap<>();
               map.put("shop_name", shops[i].getShop_name());
               map.put("shop_address", shops[i].getShop_address());
               map.put("shop_tel", shops[i].getTel());
               String picName=shops[i].getImg_name();
               int picId = getResources ( ). getIdentifier ( picName, " drawable ",
               FavoriteActivity.this.getPackageName());
               map.put("img_id", picId+"");
               data.add(map);
            }
            MyAdapter adapter=new MyAdapter
                 (FavoriteActivity.this, data, R.layout.list_item_custom,
                     new String[]{"shop_name", "shop_address", "shop_tel", "img_id"},
                     new int[]{R.id.txtName, R.id.txtAddress, R.id.txtTel, R.id.img});
            list.setAdapter(adapter);
            list.setOnItemClickListener(new OnItemClickListener(){
               @Override
               public void onItemClick(AdapterView<?>arg0, View arg1, int position,
                         long arg3){
                  Intent intent=new Intent();
                  intent.setClass(FavoriteActivity.this, ShopDetailActivity.class);
                  Bundle bundle=new Bundle();
                  bundle.putSerializable("shop", shops[position]);
                  intent.putExtras(bundle);
                  startActivity(intent);
               }
            });
         }
         class MyAdapter extends SimpleAdapter {
            public MyAdapter(Context context, List<? extends Map<String, ?>>data,
                     int resource, String[] from, int[] to){
               super(context, data, resource, from, to);
            }
            @Override
            public View getView(int position, View convertView,
                     ViewGroup parent){
               View result=super.getView(position, convertView, parent);
               TextView txtTilte=(TextView)result.findViewById(R.id.txtName);
               if(position %2 ==1){
                  result.setBackgroundColor(Color.GREEN);
                  txtTilte.setTextColor(Color.BLUE);
```

```
            } else {
                result.setBackgroundColor(Color.YELLOW);
                txtTilte.setTextColor(Color.RED);
            }
            return result;
        }
    }
    //初始化方法中,为数据库路径赋值
    private void init(Context context){
    //获取当前应用程序的包名
        String packageName=context.getPackageName();
        try {
    //获取 Application 中的元数据信息
            applicationInfo= context. getPackageManager ( ). getApplicationInfo
            (packageName, PackageManager.GET_META_DATA);
            String dbDir=applicationInfo.dataDir+File.separator+DB_DIR;
            File file=new File(dbDir);
            if(!file.exists()){
                file.mkdir();
            }
            databasePath= applicationInfo. dataDir + File. separator + DB_DIR + File.
            separator+DB_NAME;
            } catch(NameNotFoundException e){
        }
    }

}
```

店铺的实体类定义如下。

```
import java.io.Serializable;
public class Shop implements Serializable{
    private String shop_id;
    private String shop_name;
    private String shop_address;
    private String tel;
    private String img_name;
    private int img_id;
    public String getImg_name(){
        return img_name;
    }
    public void setImg_name(String img_name){
        this.img_name=img_name;
    }
    public int getImg_id(){
```

```java
        return img_id;
    }
    public void setImg_id(int img_id){
        this.img_id=img_id;
    }
    public String getShop_id(){
        return shop_id;
    }
    public void setShop_id(String shop_id){
        this.shop_id=shop_id;
    }
    public String getShop_name(){
        return shop_name;
    }
    public void setShop_name(String shop_name){
        this.shop_name=shop_name;
    }
    public String getShop_address(){
        return shop_address;
    }
    public void setShop_address(String shop_address){
        this.shop_address=shop_address;
    }
    public String getTel(){
        return tel;
    }
    public void setTel(String tel){
        this.tel=tel;
    }


}
```

店铺详情页面的 Activity 代码如下。

```java
import android.app.Activity;
import android.content.Intent;
import android.graphics.Color;
import android.os.Bundle;
import android.widget.ImageView;
import android.widget.TextView;
import cn.edu.neusoft.software.mobile.activitydemo.R;
public class ShopDetailActivity extends Activity {
private TextView txtInfo;
private Shop shop;
private ImageView img;
```

```java
@Override
protected void onCreate(Bundle savedInstanceState){
    super.onCreate(savedInstanceState);
    setContentView(R.layout.activity_shop_detail);
    txtInfo=(TextView)findViewById(R.id.txtdetail);
    img=(ImageView)findViewById(R.id.image);
    Intent intent=this.getIntent();
      shop=(Shop)intent.getSerializableExtra("shop");
      txtInfo.setTextColor(Color.RED);
      txtInfo.setTextSize(20);
    txtInfo.setText("商铺名称"+shop.getShop_name()+"\n"
         +"商铺地址"+shop.getShop_address()+"\n"+
         "商铺电话"+shop.getTel());
    int picId=getResources().getIdentifier(shop.getImg_name(),"drawable",
    ShopDetailActivity.this.getPackageName());
    img.setBackgroundResource(picId);
  }
}
```

对数据库进行操作的封装类如下。

```java
import android.content.ContentValues;
import android.database.Cursor;
import android.database.sqlite.SQLiteDatabase;
import android.database.sqlite.SQLiteException;
public class DBUtil {
  private final static String DB_NAME="coffeeshop";
  private final static int DB_VERSION=1;
  private final static String TABLE_SHOP="shop";
  public static final String KEY_ID="shop_id";
  public static final String KEY_NAME="shop_name";
  public static final String KEY_ADDRESS="shop_address";
  public static final String KEY_TELEPHONE="tel";
  public static final String KEY_IMG_NAME="img_name";
  private String databasePath;
  private SQLiteDatabase database;
  private static DBUtil dbUtil;
  private DBUtil(String databasePath){
  }
  public static DBUtil getInstance(String databasePath){

    if(dbUtil ==null){

      dbUtil=new DBUtil(databasePath);
    }
```

```java
        dbUtil.databasePath=databasePath;
        return dbUtil;
    }
    public int openDB(){

        try {
            if(database ==null || !database.isOpen()){
                database= SQLiteDatabase. openDatabase (this. databasePath, null,
                    SQLiteDatabase.OPEN_READWRITE);
            }
        } catch(SQLiteException e){
            return -1;
        }
        return 0;
    }
    public void closeDB(){
        if(database !=null && database.isOpen()){

            database.close();
            database=null;
        }

    }
    public long deleteOneData(String id){
        return database.delete(TABLE_SHOP,  "shop_id"+"=" +"'"+id+"'", null);
    }
    public Shop[] queryAllShop(){
        Cursor results=database.query(TABLE_SHOP, null,
            null, null, null, null, null);
        return ConvertToShop(results);
    }
    private Shop[] ConvertToShop(Cursor cursor){
        int resultCounts=cursor.getCount();
        if(resultCounts ==0 || !cursor.moveToFirst()){
            return null;
        }
        Shop[] shops=new Shop[resultCounts];
        for(int i=0 ; i<resultCounts; i++){
            shops[i]=new Shop();
            shops[i].setShop_id(cursor.getString(0));
            shops[i].setShop_name(cursor.getString(cursor.getColumnIndex("shop_
            name")));
            shops[i].setShop_address(cursor.getString(2));
            shops[i].setTel(cursor.getString(3));
            shops[i].setImg_name(cursor.getString(4));
```

```java
            shops[i].setImg_id(cursor.getInt(5));
            cursor.moveToNext();
        }
        return shops;
    }
    public  Cursor getShopLike(Shop shop){
        String sql="select * from "+TABLE_SHOP +" where shop_name like '%"+shop.
        getShop_name()+"%'";

        Cursor results=database.rawQuery(sql, null);
        return results;
}
public   Cursor getShop(Shop shop){
    Cursor cursor=null;
    if(shop !=null){
        return database.query(TABLE_SHOP, null, "shop_id=?"+" and "+"shop_name
        =?", new String[]{shop.getShop_id(),shop.getShop_name()}, null, null,
        null);
    }
    return cursor;
}
public Cursor getAllShop(){
    return database.query(TABLE_SHOP, null,null,null,null,null,null);
}
public long insert(Shop shop){
    ContentValues newValues=new ContentValues();
    newValues.put(KEY_ID, shop.getShop_id());
    newValues.put(KEY_NAME, shop.getShop_name());
    newValues.put(KEY_ADDRESS, shop.getShop_address());
    newValues.put(KEY_TELEPHONE, shop.getTel());
    newValues.put(KEY_IMG_NAME, shop.getImg_name());
    return database.insert(TABLE_SHOP, null, newValues);
    }
    public long deleteAllData(){
        return database.delete(TABLE_SHOP, null, null);
    }
public long updateOneData(String id ,Shop shop){
    ContentValues updateValues=new ContentValues();
    updateValues.put(KEY_NAME, shop.getShop_name());
    updateValues.put(KEY_ADDRESS, shop.getShop_address());
    updateValues.put(KEY_TELEPHONE, shop.getTel());
    updateValues.put(KEY_IMG_NAME, shop.getImg_name());
    return database.update(TABLE_SHOP, updateValues,  KEY_ID+"=" +"'"+id+
    "'", null);
```

        }
    }

shop_layout.xml 文件的内容如下。

```xml
<?xml version="1.0" encoding="utf-8"?>
<LinearLayout xmlns:android="http://schemas.android.com/apk/res/android"
    android:layout_width="match_parent"
    android:layout_height="match_parent"
    android:orientation="vertical"
    android:background="#FFFFB9">
<ListView
        android:id="@+id/listshop"
        android:layout_width="match_parent"
        android:layout_height="wrap_content" >
</ListView>
</LinearLayout>
```

list_item_custom.xml 文件的内容如下。

```xml
<?xml version="1.0" encoding="utf-8"?>
<LinearLayout xmlns:android="http://schemas.android.com/apk/res/android"
    android:layout_width="match_parent"
    android:layout_height="match_parent"
    android:orientation="horizontal" >
<ImageView
        android:id="@+id/img"
        android:layout_width="100dp"
        android:layout_height="100dp"
        android:src="@drawable/a1"
         />
<LinearLayout
        android:layout_width="wrap_content"
        android:layout_height="wrap_content"
        android:orientation="vertical"
        android:layout_marginLeft="15dp" >
<TextView
        android:id="@+id/txtName"
        android:layout_width="wrap_content"
        android:layout_height="wrap_content"
        android:textColor="#f00"
        android:textStyle="bold"
        android:textSize="18sp"
        android:text="dssd"/>
<TextView
        android:id="@+id/txtAddress"
```

```xml
        android:layout_width="wrap_content"
        android:layout_height="wrap_content"
        android:textColor="#00f"
        android:textStyle="bold"
        android:textSize="14sp"
        android:text="dssd"/>
<TextView
        android:id="@+id/txtTel"
        android:layout_width="wrap_content"
        android:layout_height="wrap_content"
        android:textColor="#00f"
        android:textStyle="bold"
        android:textSize="14sp"
        android:text="dssd"/>
</LinearLayout>

</LinearLayout>
```

activity_shop_detail.xml 文件的内容如下。

```xml
<RelativeLayout xmlns:android="http://schemas.android.com/apk/res/android"
    xmlns:tools="http://schemas.android.com/tools"
    android:layout_width="match_parent"
    android:layout_height="match_parent"
tools:context=".ShopDetailActivity" >
<TextView
        android:layout_width="wrap_content"
        android:layout_height="wrap_content"
        android:text="@string/hello_world"
        android:id="@+id/txtdetail"
      />
<ImageView
        android:id="@+id/image"
        android:layout_width="wrap_content"
        android:layout_height="wrap_content"
        android:layout_below="@+id/txtdetail"
        android:layout_centerInParent="true"
/>
<Button
        android:id="@+id/button1"
        android:layout_width="fill_parent"
        android:layout_height="wrap_content"
        android:layout_alignParentBottom="true"
```

```xml
        android:text="加入收藏夹" />
</RelativeLayout>
```

11.6.3 项目说明

在 AndroidManifest.xml 中，<meta-data>元素可以作为子元素，被包含在<activity>、<application>、<service>和<receiver>元素中。不同的父元素，应用时读取的方法也不同。

(1) 在 Activity 中应用<meta-data>元素。

XML 代码段如下。

```xml
<activity...>
    <meta-data android:name="data_Name" android:value="hello my activity">
    </meta-data>
</activity>
```

Java 代码段如下。

```java
ActivityInfo info=this.getPackageManager()
              .getActivityInfo(getComponentName(),
              PackageManager.GET_META_DATA);
String msg =info.metaData.getString("data_Name");
Log.d(TAG, " msg =="+msg);
```

(2) 在 application 中应用<meta-data>元素。

XML 代码段如下。

```xml
<application...>
    <meta-data android:value="hello my application" android:name="data_Name">
    </meta-data>
</application>
```

Java 代码段如下。

```java
ApplicationInfo appInfo=this.getPackageManager()
                   .getApplicationInfo(getPackageName(),
              PackageManager.GET_META_DATA);
String msg=appInfo.metaData.getString("data_Name");
Log.d(TAG, " msg =="+msg);
```

(3) 在 service 中应用<meta-data>元素。

XML 代码段如下。

```xml
<service android:name="MetaDataService">
    <meta-data android:value="hello my service" android:name="data_Name">
    </meta-data>
</service>
```

Java 代码段如下。

```
ComponentName cn=new ComponentName(this, MetaDataService.class);
ServiceInfo info=this.getPackageManager()
                .getServiceInfo(cn, PackageManager.GET_META_DATA);
String msg=info.metaData.getString("data_Name");
Log.d(TAG, " msg =="+msg);
```

（4）在 receiver 中应用＜meta-data＞元素。

XML 代码段如下。

```
<receiver android:name="MetaDataReceiver">
    <meta-data android:value="hello my receiver" android:name="data_Name">
    </meta-data>
        <intent-filter>
            <action android:name="android.intent.action.PHONE_STATE"></action>
        </intent-filter>
</receiver>
```

Java 代码段如下。

```
ComponentName cn=new ComponentName(context, MetaDataReceiver.class);
ActivityInfo info=context.getPackageManager()
                .getReceiverInfo(cn, PackageManager.GET_META_DATA);
String msg=info.metaData.getString("data_Name");
Log.d(TAG, " msg =="+msg);
```

本 章 小 结

Android 的数据存储方式主要有以下几种。

使用 SharedPreferences 存储数据：最适合 SharedPreferences 的地方就是保存配置信息。

Internal Storage 内部存储空间：与其他（外部的）存储相比，有着比较稳定、存储方便、操作简单、更加安全（因为可以控制访问权限）等优点。唯一的缺点就是存储空间有限。

External Storage 外部存储空间：存储不稳定。

SQLite 数据库存储数据：存储结构化数据。

这几种方式各有各的优点和缺点，要根据实际情况来选择，而无法给出统一的标准。对比这几种方式，可以总结如下。

（1）简单数据和配置信息，SharedPreference 是首选。

（2）如果 SharedPreferences 不够用，就创建一个数据库。

（3）对于结构化数据，一定要创建数据库，虽然这稍显烦琐，但是好处无穷。

（4）Android 的文件系统就是用来存储文件（也即非配置信息或结构化数据）的，如文本文件、二进制文件、PC 文件、多媒体文件、下载的文件等。

（5）尽量不要创建文件。

（6）如果创建的是私密文件或是重要文件，就存储在内部，否则放到外部存储。

（7）若数据量较大且需要服务器帮助处理数据，就需要用到网络存储。

本章习题

1．Android 中常用的数据存储方式有哪些？怎样去实现这些存储方式？
2．使用 SharedPrefernce 的基本步骤有哪些？
3．文件存储常用输入输出流有哪些？
4．创建名为 test.db 的数据库，并建立 staff 数据表，表内的属性值如表 11.4 所示。

表 11.4 staff 数据表

属　　性	数据类型	说　　明
_id	integer	主键
name	text	姓名
sex	text	性别
department	text	所在部门
salary	float	工资

5．设计 CoffeeStore 项目购物车数据库表的结构，并在 coffee.db 数据库中添加购物车表 shop.db，编写对 shop.db 进行增删改查操作的方法。

6．SharedPreferences 类实现数据的存储和读取需要使用的内部接口名是什么？说明项目中用到 SharedPreferences 的位置。

网络存储技术

本章概述

本章讲解 Android 的异步任务使用、JSON 数据格式解析和网络编程技术,包括自定义异步任务类、JSON 数据类型、解析 JSON 对象的方法、Android 常见的 HTTP 通信接口、HttpURLConnection 的常见方法使用、利用 HttpURLConnection 完成与服务器端数据的接收和发送等基本操作。

学习重点与难点

重点:

(1) 异步任务类 AsyncTask。

(2) JSON 格式的解析。

(3) HttpURLConnection 网络通信接口。

难点:

(1) 编写自定义的异步任务类。

(2) 使用 HttpURLConnection 接口的常用方法。

学习建议

编写一些简单程序实例,体会异步任务的作用和编程流程。查阅网上资料,了解 HTTP 通信原理和过程,理解 HTTP 请求和响应数据格式,理解网络通信程序的代码含义。通过大量的客户端—服务端编程实例的练习,巩固知识点和强化记忆,提高熟练运用 HttpURLConnection 接口常用方法的能力。

12.1 异步任务

12.1.1 异步任务的使用场合

Android 应用程序完全运行在一个独立的线程中,通常称为主线程(或 UI 线程)。Android 默认约定主线程阻塞超过 20 秒时,将会发 ANR(Application Not Responding)异常。但实际上,当应用程序等待超过 5 秒,用户都会感到无法容忍。当出现输入事件(如按键、触屏事件)的响应超过 5 秒或者意图接收器(intentReceiver)超过 10 秒钟仍未执行完毕的情况,就会显示 ANR。因此,不要在主线程中执行任何耗时的操作,而将耗时操作放在子线程中执行。由于 Android 的 UI 操作不是线程安全的,所以要求所有更新 UI 的操作必须在主线程中执行。基于以上因素,需要提供一个主线程和子线程之间通信的机制,而异步任务可以解决这一问题。

Android 提供了一个轻量级、适用于简单的异步处理的类,即 AsyncTask 类。使用这个类不需要借助线程和 Handler 即可完成线程间的通信。

12.1.2 异步任务类

异步任务类 AsyncTask<Params,Progress,Result>是个抽象类。在自定义异步任务时,继承这个抽象类即可,继承时需要设定 3 个泛型参数,内容如下。

(1) params 是指调用 execute()方法启动异步任务时传入的参数类型和 doInBackgound()的参数类型。

(2) progress 是指更新进度时传递的参数类型,即 publishProgress() 和 onProgressUpdate()的参数类型。

(3) result 是指 doInBackground()的返回值类型。

自定义异步任务的步骤如下。

(1) 创建 AsyncTask<Params,Progress,Result>的子类,并为 3 个泛型参数指定具体的类型。如果某个泛型参数不需要指定类型,可将它指定为 Void 类型。

(2) 根据需要,实现 AsyncTask 的方法如下。

① doInBackground(Params ...):运行于子线程中,在此方法中进行后台线程将要完成的任务,可以调用 publishProgress(Progress... values)方法更新任务执行进度。

② onProgressUpdate(Progress... values):运行于主线程中,在 doInBackground()方法中调用 publishProgress()方法更新任务的执行进度后,就会触发该方法。

③ onPreExecute():运行于主线程中,执行后台耗时操作前被调用,通常可以在这个方法中完成一些初始化操作,比如在界面上显示进度条。

④ onPostExecute(Result result):doInBackground()执行完毕,系统会自动调用 onPostExecute()方法,并将 doInBackground()方法的返回值传给该方法的参数 result。

(3) 调用 AsyncTask 子类对象的 execute(Params... params)方法开始执行异步任务。

使用 AsyncTask 类时,需要遵守以下几条准则。

① 必须在主线程中创建异步任务的对象。

② 必须在主线程中调用启动异步任务的方法 execute()。

③ 不要手动调用 onPreExecute()、onPostExecute()、doInBackground()、onProgressUpdate()这些方法。

④ 异步任务只能被执行一次,多次调用时将会出现异常。

【例 12-1】 使用 AsyncTask 实现简易计数器。

创建 AsyncTaskActivity,实现一个简易的计数功能,单击 Start 按钮开始计数,并将计数值显示在 TextView 控件上。单击 Stop 按钮停止计数。

在 AsyncTaskActivity 中声明全局变量 isRun,初值为 true,用于记录是否停止计数,代码如下。

```
private boolean isRun=true;
```

AsyncTaskActivity 的 onCreate()方法如下。

```
protected void onCreate(Bundle savedInstanceState){
super.onCreate(savedInstanceState);
setContentView(R.layout.activity_asyn_task);
textView= (TextView)findViewById(R.id.textView2);
Button button= (Button)findViewById(R.id.button2);
button.setOnClickListener(new View.OnClickListener(){
        @Override
        public void onClick(View v){
            new MyTask().execute();
        }
});
Button button= (Button)findViewById(R.id.button2);
button2.setOnClickListener(new View.OnClickListener(){
@Override
        public void onClick(View v){
            isRun=false;
        }
});
}
```

在新建的 AsyncTaskActivity 的 onCreate()方法中添加一个 TextView 和两个 Button，编写 Start 按钮的单击事件处理方法，调用自定义的异步任务类 MyTask 的 execute()方法，启动一个异步任务，在异步任务中执行计数器增 1 的操作。编写 Stop 按钮的单击事件处理方法，自定义的异步任务类 MyTask 代码如下。

```
class MyTask extends AsyncTask<String,String,Integer>{
private String text;
//1.在 UI 线程执行
@Override
protected void onPreExecute(){
    super.onPreExecute();
    text=textView.getText().toString();
    System.out.println("onPreExecute");
}
//2.在子线程执行
//String...:可变参数,个数从 0 个到 n 个,要求所有参数的类型一致
protected Integer doInBackground(String... arg0){
    //TODO Auto-generated method stub
    System.out.println("doInBackground");
    int i=0;
```

```
        while(isRun){
            i=Integer.parseInt(text);
            i++;
            //发布进度
            publishProgress(i+"");
            try {
           .Thread.sleep(500);
            } catch(InterruptedException e){
            e.printStackTrace();
            }
        }
        return i;
}
//3.在 UI 线程执行
@Override
protected void onProgressUpdate(String... values){
    //TODO Auto-generated method stub
    super.onProgressUpdate(values);
    System.out.println("onProgressUpdate");
    textView.setText(values[0]);
    text=textView.getText().toString();
}
//4.UI 线程执行
@Override
protected void onPostExecute(Integer result){//
    //TODO Auto-generated method stub
    super.onPostExecute(result);
    System.out.println("onPostExecute");
    System.out.println("i:"+result);         //可访问到异步任务结束后 i 的值
    isRun=true;
}
}
```

启动异步任务后,首先执行 onPreExecute()方法,进行一些初始化操作。在 onPreExecute()方法中获取 TextView 的初始显示内容。

接下来执行 doInBackground()方法,在子线程中执行耗时任务。在这个方法中,若 isRun 值为 true,则每隔半秒钟计数值增 1,并调用 publishProgress()方法更新 TextView,显示当前的计数值。

12.2 JSON 数据解析

12.2.1 JSON 简介

考虑到 XML 结构的复杂程度以及解析它所需的代价,在普通的网络应用开发中,无论是服务器端生成或处理 XML,还是客户端程序解析 XML,都经常需要编写复杂的代码,造成极低的开发效率。而 JSON 的出现则可以解决这些方面的问题。与 XML 相比,JSON 没有结束标签,并且更加简短,读写的速度更快,解析起来也比 XML 容易,Android SDK 中有现成的工具类对其直接进行解析。JSON 支持数组结构,并且没有使用任何保留字。

JSON(Javascript Object Notation)是一种轻量级的数据交换格式,易于阅读和编写,也易于机器解析和生成。和 XML 一样,JSON 也是基于纯文本的数据格式,它是一系列键值对的集合,比 XML 更小、更快,更易解析。JSON 的数据格式非常简单,可以用 JSON 传输一个简单的字符串、数值或布尔值,也可以传输一个数组或者一个复杂的 Object 对象。

12.2.2　JSON 的基本语法

JSON 中支持的基本数据类型如下。
- 数字(整数或浮点数)
- 字符串(在双引号中)
- 逻辑值(true 或 false)
- 数组(在方括号中)
- 对象(在花括号中)
- null

JSON 中的数据结构主要包括两种：对象和数组,通过这两种结构可以表示各种复杂的数据结构。

(1) 对象：JSON 对象的书写格式是"名/值"对,名和值之间使用":"隔开,每个"名/值"对之间使用","分隔,并且使用"{"和"}"括起来,一般的形式如下。

```
{name1:value1, name2:value2,..., nameN:valueN}
```

通常会使用"对象名.name"的方式来获取对应的属性值 value。这个属性值的类型可以是数字、字符串、对象或数组。

例如,为了描述一个 User 对象,其中包括用户的 firstName 和 lastName 属性,可以使用如下的 JSON 对象。

```
String jsonStr="{\"firstName\":\"John\" ,\"lastName\":\"Doe\" }";
```

在 Android 中,可以使用 org.json 工具包中的 JSONObject 来描述以上 JSON 对象,例如：

```
JSONObject jo=new JSONObject(jsonStr);
```

(2) 数组：JSON 数组相当于值的有序列表,一个或者多个值直接使用","分隔,并且使用"["和"]"括起来,一般的形式如下。

```
[value1, value2,..., valueN]
```

通常会使用"数组名[index]"的方式获取对应的值,值的类型可以是数字、字符串、对象或数组。

例如,如果需要描述一个 User 对象列表,而不是单一的一个 User 对象,可以使用如下的 JSON 数组:

```
String jsonStr="[{\"firstName\":\"John\" ,\"lastName\":\"Doe\" }, { \"firstName\":\"Anna\" , \"lastName\":\"Smith\" }, { \"firstName\":\"Peter\" , \"lastName\":\"Jones\" } ]";
```

在 Android 中,可以使用 org.json 工具包中的 JSONArray 来描述以上 JSON 数组,例如:

```
JSONArray ja=new JSONArray(jsonStr);
```

12.2.3 JSON 的解析

Android SDK 提供了 org.json 工具包,可以用来解析 JSON。常用方法如下。

- Object opt(String name)
- int optInt(String name)
- long optLong(String name)
- boolean optBoolean(String name)
- float optFloat(String name)
- double optDouble(String name)
- JSONArray optJSONArray(String name)
- JSONObject optJSONObject(String name)
- Object get(String name)
- int getInt(String name)
- long getLong(String name)
- boolean getBoolean(String name)
- float getFloat(String name)
- double getDouble(String name)
- JSONArray getJSONArray(String name)
- JSONObject getJSONObject(String name)
- booleanhas(String name)

建议使用 opt 开头的方法,因为这些方法在解析时,如果对应属性不存在,会返回空值或者 0,不会报异常。而使用 get 开头的方法,则不会判断是否存在该属性,需要调用 has()方法判断,如果属性不存在,会抛出异常。

下面以实际例子说明在 Android 中解析 JSON 格式数据的方法。

假设服务端返回的 JSON 格式的响应数据如下。

```
{
    "pageCount":1,
```

```
    "list":[{
        "id":1,
        "name":{"firstName":"san","lastName":"zhang"},
        "age":"男",
        "email":"zs@sina.cn"
    },{
      "id":2,
        "name":{"firstName":"ming","lastName":"li"},
        "age":"男",
        "email":"lm@163.com"
    },{
        ...
    }]
}
```

在Android中,可以使用如下代码进行解析。

```
public void parse(JSONObject jo){
    if(jo !=null){
        try {
            JSONArray jsonArray=jo.optJSONArray("list");
            if(jsonArray !=null){
                int length=jsonArray.length();
                if(0<length){
                    ArrayList<Map<String,Object>>date=new ArrayList<>();
                    for(int i=0; i<length; i++){
                        JSONObject jsonObj=jsonArray.optJSONObject(i);
                        if(jsonObj ==null)
                            continue;
                        Map<String, Object>m=new HashMap<String, Object>();
                        m.put("id", jsonObj.optString("id"));
                        m.put("name", jsonObj.optJSONObject("name"));
                        m.put("age", jsonObj.optInt("age"));
                        m.put("email", jsonObj.optString("email"));
                        data.add(m);
                    }
                    Log.v("data", data.toString());
                }
            }
        } catch(JSONException e){
            //TODO Auto-generated catch block
            e.printStackTrace();
        }
    }
}
```

12.3 HttpURLConnection

12.3.1 HTTP 通信接口

Android 开发中常见的 HTTP 通信接口有以下两种。

1) 标准 Java 网络接口(java.net)

java.net 包中提供了与网络通信有关的类和接口,可以创建 URL 对象,以及 URLConnection、HttpURLConnection 对象,设置链接参数、向服务器读/写数据等。

2) Apache 网络接口(org.apache.http)

Apache Httpclient 是一个开源项目,功能较为完善,可以发送 HTTP 请求,接受 HTTP 响应,为客户端的 HTTP 编程提供高效、功能丰富的工具包支持,但很难在不影响其兼容性的情况下对其进行改进。从 Android 6.0 之后,谷歌已经移除了对 Apache HTTP client 的支持,并建议使用 HttpURLConnection 来代替。

12.3.2 HttpURLConnection 的常用方法

使用 HttpURLConnection 进行网络通信,常用的有以下几个方法。

1) 创建一个 URL 对象

```
URL url=new URL("http://www.baidu.com");
```

2) 调用 URL 对象的 openConnection()方法创建 HttpURLConnection 对象

```
HttpURLConnection conn=(HttpURLConnection)url.openConnection();
```

3) 设置连接超时时间

```
conn.setConnectTimeout(3000);
```

4) URL 连接可用于输入或输出,可使用 setDoInput()方法设置使用 URL 连接输入,使用 setDoOutput()方法设置使用 URL 连接输出

```
conn.setDoInput(true);
conn.setDoOutput(true);
```

5) 设置请求方式

```
conn.setRequestMethod("POST");
```

6) 设置是否使用缓存

```
conn.setUseCaches(false);
```

7) 设置 HTTP 请求头属性

```
conn.setRequestProperty("Connection", "Keep-Alive");    //维持长连接
conn.setRequestProperty("Charset", "UTF-8");            //设置字符集
```

```
conn.setRequestProperty("Content-Type","application/x-www-form-
urlencoded");                                                    //设置响应类型
```

8) 建立 URL 连接

```
conn.connect();
```

9) 获取输出流对象，getOutputStream()方法会隐含地进行 connect()方法调用

```
DataOutputStream out=new DataOutputStream(conn.getOutputStream());
```

10) 利用 URLEncoder 或 UrlEncodedFormEntity，对发送到服务器端的数据进行编码

```
String content =URLEncoder.encode(data, HTTP.UTF_8);
```

或者

```
UrlEncodedFormEntity uf=new UrlEncodedFormEntity(data,HTTP.UTF_8);
```

11) 向服务器发送数据

```
out.writeBytes(content);
out.flush();
out.close();
```

或者

```
uf.writeTo(out);
out.flush();
out.close();
```

12) 对服务器端返回的响应码进行判断，判断响应码是否为 HttpURLConnection.HTTP_OK(HttpURLConnection.HTTP_OK 值为 200，代表响应成功)

```
if(conn.getResponseCode()==HttpURLConnection.HTTP_OK)
```

13) 获取输入流对象

```
InputStream in=conn.getInputStream();
```

14) 读取数据

```
String result=readData(is, "GBK");
```

或者

```
BufferedReader br=new BufferedReader(new InputStreamReader(in));
String result=br.readLine();
```

15) 断开连接

```
conn.disconnect();
```

12.4 利用异步任务读取服务器端图片信息

由于客户端程序与服务器端程序进行网络通信属于典型的耗时操作,所以需要利用前面介绍的异步任务来完成网络连接的建立、服务器端数据读取和发送的操作。

本节介绍如何在异步任务中建立与服务器端的通信,读取服务器端传回的图片信息。

在 TestAsyncGetImageActivity 中放置一个 ImageView 控件,用于显示服务器端传回的图片,声明全局变量 imageView。

```
private ImageView imageView;
```

在 onCreate()方法里启动异步任务,代码如下。

```
@Override
protected void onCreate(Bundle savedInstanceState){
    super.onCreate(savedInstanceState);
    setContentView(R.layout.activity_main);
    imageView= (ImageView)findViewById(R.id.imageView1);
    AsynGetImage agi=new AsynGetImage();
    agi.execute(new URL("http://localhost:8080/test/img1.jpg"));
}
```

编写异步任务类,用于执行网络通信、读取图片的操作,代码如下。

```
private class AsyncGetImage extends AsyncTask<URL, Void, Bitmap>{
protected Bitmap doInBackground(URL... params){
    Bitmap result=mGetDataFromServer(params[0]);
    return result;
}
private Bitmap mGetDataFromServer(URL url){
    Bitmap bitmap=null;
    HttpURLConnection conn=null;
    try{
      conn= (HttpURLConnection)url.openConnection();
      conn.setConnectTimeout(3000);
      conn.setDoInput(true);
      conn.connect();
      if(HttpURLConnection.HTTP_OK ==conn.getResponseCode()){
          InputStream is=conn.getInputStream();
          bitmap=BitmapFactory.decodeStream(is);
          is.close();
      }
    } catch(IOException e){
      e.printStackTrace();
      Log.v("AsynGetImage:", e.getMessage());
```

```
        } finally {
           conn.disconnect();
        }
        return bitmap;
    }
    @Override
    protected void onPostExecute(Bitmap result){
        if(result!=null){
    imageView.setImageBitmap(result);
        }
    }
}
```

12.5 项目实战：登录功能

12.5.1 项目分析

项目启动后，需要输入合法的用户名和密码登录。也就是说，需要将用户在 Android 客户端程序界面中输入的用户名、密码等信息传送到服务器端程序，服务端程序进行判断，返回是否合法的结果。Android 客户端程序接收到服务端返回结果后，解析结果数据，根据结果给予用户不同的响应，如果是合法用户，则登录成功，进入 CoffeeStore 主页面，否则，登录失败，提示错误信息。

12.5.2 项目实现

首先创建 2 个 Activity，分别表示项目登录页和登录后主页，2 个 Activity 名称如下。

在 LoginActivity 中使用 HttpURLConnection 进行与服务器端程序的网络通信，解析服务器端返回的 JSON 数据。若用户名和密码正确，则登录成功，跳转到主页 MainActivity，否则提示"用户名或密码错误"。

LoginActivity 的代码如下。

```java
public class LoginActivity extends Activity implements IAsynHttpCallBack{
private EditText etUsername;
    private EditText etPassword;
@Override
protected void onCreate(Bundle savedInstanceState){
        super.onCreate(savedInstanceState);
        setTheme(R.style.CustomTheme);
        requestWindowFeature(Window.FEATURE_CUSTOM_TITLE);
        setContentView(R.layout.activity_login);
```

```java
        getWindow().setFeatureInt(Window.FEATURE_CUSTOM_TITLE,R.layout.
        titlebar);
        etUsername=(EditText)findViewById(R.id.et_username);
        etPassword=(EditText)findViewById(R.id.et_password);
        //实例化登录页面的监听器对象
        LoginOnClickListener loginOnClickListener=new LoginOnClickListener();
        //为登录按钮绑定监听器
        Button btnLogin=(Button)findViewById(R.id.btn_login);
        btnLogin.setOnClickListener(loginOnClickListener);
    }
    //登录按钮单击事件监听器
    class LoginOnClickListener implements View.OnClickListener{
        @Override
        public void onClick(View v){
            if(v.getId()==R.id.btn_login){                    //登录按钮被单击
                final String userName=etUsername.getText().toString().trim();
                final String password=etPassword.getText().toString().trim();
                if(userName.equals("")||password.equals("")){   //用户名或者密码未填写
                    Toast.makeText(getApplicationContext(),"请将用户名密码填写完全后
                    再登录",Toast.LENGTH_LONG).show();
                }
                else{
//创建网络通信工具类对象,与服务端登录程序通信
                    AsynHttp ao=new AsynHttp("http://10.0.2.2:8080/
                    CoffeeStoreServer/login.action",LoginActivity.this,
                    LoginActivity.this);
//将用户名和密码作为请求参数
List<NameValuePair>data=new ArrayList<NameValuePair>();
data.add(new BasicNameValuePair("username", userName));
data.add(new BasicNameValuePair("pwd", password));
ao.set_data(data);
//调用工具类的execute()方法,启动工具类中自定义的异步任务
ao.execute();
                }
            }
        }
    }
    //网络通信完毕,服务器端返回响应后的回调函数
    @Override
    public void callBackFunction(JSONObject jsonobj){
        if(jsonobj!=null){
        try {
            //解析服务器端返回的JSON数据
            JSONObject j=jsonobj.optJSONObject("loginReturn");
```

```
            if(j.optString("loginFlag").equals("1")){        //登录成功
                Intent intent=new Intent(this,MainActivity.class);
startActivity(intent);
                this.finish();                               //关闭当前登录窗口
            }
            else{                                            //登录失败
                Toast.makeText(this, j.getString("msg"),Toast.LENGTH_SHORT).show();
        }
        } catch(JSONException e){
            //TODO Auto-generated catch block
            e.printStackTrace();
        }
        }
    }
}
```

由于项目中有多个功能都需要进行与服务器端的网络通信,而大部分网络通信代码基本相同,所以可以将完成网络通信功能的异步任务等代码封装起来,放在一个通用的工具类 AsynHttp 中来完成。这样可以有效减少项目中的冗余代码,使程序结构更加清晰。工具类 AsynHttp 代码如下。

```
    public class AsynHttp {
    private String _url=null;
        private List<NameValuePair> _data=null;
        private Context _context=null;
        private IAsynHttpCallBack _callbackFunction=null;
        private AsyncTaskObjectClass _asynTaskObject=null;
        private boolean accessServer=true,getJSON=true,formedUrl=true;
        private boolean _getCookie=false;
        public AsynHttp(){
_asynTaskObject=new AsyncTaskObjectClass();
_getCookie=false;
}
    public AsynHttp(String url, Context context, IAsynHttpCallBack callback){
_url=url;
_context=context;
_callbackFunction=callback;
_asynTaskObject=new AsyncTaskObjectClass();
_getCookie=false;
}
    public AsynHttp(String url, Context context, IAsynHttpCallBack callback,boolean getCookie){
_url=url;
_context=context;
```

```java
_callbackFunction=callback;
_asynTaskObject=new AsyncTaskObjectClass();
_getCookie=getCookie;
}
public AsynHttp(String url, List<NameValuePair> data, Context context,
IAsynHttpCallBack callback){
_url=url;
_data=data;
_context=context;
_callbackFunction=callback;
_getCookie=false;
_asynTaskObject=new AsyncTaskObjectClass();
}
public AsynHttp(String url, List<NameValuePair> data, Context context,
IAsynHttpCallBack callback,boolean getCookie){
_url=url;
_data=data;
_context=context;
_callbackFunction=callback;
_getCookie=getCookie;
_asynTaskObject=new AsyncTaskObjectClass();
}
public void set_data(List<NameValuePair> _data){
this._data=_data;
}
public void set_asynTaskObject(AsyncTaskObjectClass _asynTaskObject){
this._asynTaskObject=_asynTaskObject;
}
public void set_context(Context _context){
this._context=_context;
}
public void set_callbackFunction(IAsynHttpCallBack _callbackFunction){
this._callbackFunction=_callbackFunction;
}
public void set_url(String _url){
this._url=_url;
}
public void execute(){
try {
    //启动用于执行网络通信操作的异步任务
_asynTaskObject.execute(new URL(_url));
} catch(MalformedURLException e){
//TODO Auto-generated catch block
e.printStackTrace();
```

```java
                formedUrl=false;
            }
        }
        //自定义的异步任务类,主要用于执行网络通信操作
private class AsyncTaskObjectClass extends AsyncTask<URL, Void, JSONObject>{
            //异步任务后台线程,负责执行网络通信操作
protected JSONObject doInBackground(URL... params){
//TODO Auto-generated method stub
JSONObject result=mGetDataFromServer(params[0]);
returnresult;
}
private JSONObject mGetDataFromServer(URL url){
            InputStream in=null;
HttpURLConnection urlConnection=null;
JSONObject json=null;
String cookie="";
        try {
                //创建 HttpURLConnection 对象
            urlConnection=(HttpURLConnection)url.openConnection();
//设置使用 URL 连接进行输入
urlConnection.setDoInput(true);
//设置连接超时时间为 6000ms
urlConnection.setConnectTimeout(6000);
//设置请求方式为"POST"方式
urlConnection.setRequestMethod("POST");
urlConnection.setUseCaches(false);
//配置本次连接的 Content-type,配置为 application/x-www-form-urlencoded 的
urlConnection.setRequestProperty("Content-Type",
"application/x-www-form-urlencoded");
//在请求头中设置 Cookie 属性,其中包含 SESSIONID 信息
PrefStore pref=PrefStore.getInstance(_context);
        if(!_getCookie){
            urlConnection.addRequestProperty("Cookie",pref.getPref
            ("cookie", ""));
}
if(null !=_data){
//设置使用 URL 连接进行输出操作
                urlConnection.setDoOutput(true);
//获取输出流对象
DataOutputStream out=new DataOutputStream(
urlConnection.getOutputStream());
UrlEncodedFormEntity uf=new UrlEncodedFormEntity(_data,
HTTP.UTF_8);
uf.writeTo(out);
```

```
        out.flush();
        out.close();
    }
    //若响应码为200,即代表响应成功
    if(HttpURLConnection.HTTP_OK ==urlConnection
                    .getResponseCode()){
    if(!url.getHost().equals(
    urlConnection.getURL().getHost())){
    accessServer=false;
    } else {
    if(_getCookie){
            //获取响应头的所有属性
                    Map < String, List < String > > map = urlConnection.
                    getHeaderFields();
    //得到响应头中Cookie的所有内容,其中包括SESSIONID
    List<String>list=(List<String>)map.get("Set-Cookie");
                    if(list !=null && list.size()!=0){
                        StringBuilder builder=new StringBuilder();
                        for(String str : list){
                            cookie=builder.append(str).toString();
    }
                    }
    //使用SharedPreferences保存本次连接的服务端SESSIONID
                        pref.savePref("cookie", cookie);
    }
                    in=urlConnection.getInputStream();
    BufferedReader br=new BufferedReader(
    new InputStreamReader(in));
    String a=br.readLine();
    json=new JSONObject(a);
    in.close();
    }
            }
        } catch(IOException e){
    //TODO Auto-generated catch block
    e.printStackTrace();
    accessServer=false;
    } catch(JSONException e){
    //TODO Auto-generated catch block
    e.printStackTrace();
    getJSON=false;
    } finally {
            urlConnection.disconnect();
    }
```

```
            return json;
        }
        @Override
        protected void onPostExecute(JSONObject result){
            if(!formedUrl)
                    Toast.makeText (_context, ConfigureClass.FORMED_URL_ERROR, Toast.
                        LENGTH_SHORT).show();
                else if(!getJSON)
                    Toast.makeText (_context, ConfigureClass.GET_JSON_ERROR, Toast.
                        LENGTH_SHORT).show();
                else if(!accessServer)
                    Toast.makeText(_context, ConfigureClass.ACCESS_SERVER_ERROR,Toast.
                        LENGTH_SHORT).show();
                else if(result==null)
                    Toast.makeText(_context, ConfigureClass.RESPONSE_JSON_NULL,Toast.
                        LENGTH_SHORT).show();
        _callbackFunction.callBackFunction(result);
        }
    }
}
```

登录功能中涉及与服务器端登录程序的交互,服务器端程序使用 Servlet 接收客户端请求,读取请求参数(包括用户名和密码等参数),并调用 Service 层进行业务操作,连接后台数据库,查找用户信息表,判断用户名和密码是否正确,并使用 JSON 格式将判断结果返回给客户端。服务器端登录 Servlet 代码如下。

```
@WebServlet("/login.action")
public class LoginActionextends HttpServlet{
    protected void doPost(HttpServletRequest request,
            HttpServletResponse response)throws ServletException, IOException {
        //TODO Auto-generated method stub
        PrintWriter out=response.getWriter();
        HttpSession session=request.getSession();
        String username=request.getParameter("username");
        String pwd=request.getParameter("pwd");
        String result="";
        UserService us=newUserService();
        intuserid=us.checkUser(username, pwd);
        if(userid>0){                        //登录成功
            session.setAttribute("userid", userid);
            result="{\"loginReturn\":{\"loginFlag\":\"1\",\"msg\":\"\"}}";
        }
        else{                                //登录失败
            result="{\"loginReturn\":{\"loginFlag\":\"0\",\"msg\":\"用户名或密
```

```
        码错误\"}}";
    }
    out.print(result);
    out.flush();
    out.close();
}
protected void doGet (HttpServletRequest request, HttpServletResponse
response)throws ServletException, IOException {
    //TODO Auto-generated method stub
    this.doPost(request, response);
}
}
```

服务器端 Service 的代码如下。

```
public class UserService{
    public int checkUser(String username,String pwd){
        String sql="select userid from userinfo where username=? and pwd=?";
        return  DBUtil.getInt(sql, new Object[]{username,pwd});
    }
}
```

由于 Service 中的多个方法都会利用 JDBC 连接数据库，所以将访问数据库操作的代码封装起来，放在一个通用的工具类 DBUtil 中，减少代码冗余。数据库访问工具类的 DBUtil 代码如下（这里仅列出和登录功能、店铺列表功能相关的方法代码）。

```
public class DBUtil {
    private static Connection conn=null;                         //连接对象
    private static PreparedStatement pstmt=null;                 //语句对象
    private static ResultSet rs=null;                            //结果集对象
    private static String datasourceName=Const.DATA_SOURCE;      //数据源名称
    public DBUtil(){
    }
    public static Connection getConn(){
        returnconn;
    }
    public staticvoid setConn(Connection con){
        conn=con;
    }
    //获取连接对象
    public static Connection getConnection()throws NamingException,
    SQLException {
        Context ctx=null;
        DataSource ds=null;
        ctx=new InitialContext();
        ds=(DataSource)ctx.lookup(datasourceName);
```

```java
        conn=ds.getConnection();
        returnconn;
    }
    //获取语句对象
    private static PreparedStatement getPrepareStatement(String sql)
            throws NamingException, SQLException {
        if(conn==null||conn.isClosed())
            pstmt=getConnection().prepareStatement(sql);
        else
            pstmt=conn.prepareStatement(sql);
        return pstmt;
    }
    //关闭对象
    public static void close(){
        try {
            if(rs !=null)
                rs.close();
            if(pstmt !=null)
                pstmt.close();
            if(conn !=null)
                conn.close();
        } catch(SQLException e){
            e.printStackTrace();
        }
    }
    //遍历参数数组,将数组中的值按位置一一对应地对 pstmt 所代表的 SQL 语句中的参数进行设置
    private static void setParams(String sql, Object[] params)throws NamingException,
            SQLException {
        pstmt=getPrepareStatement(sql);
        for(inti=0; i<params.length; i++)
            pstmt.setObject(i+1, params[i]);
    }
    //从结果集中得到一个对象
    private static Object getObjectFromRS(String sql, Object[] params)
            throws NamingException, SQLException {
        Object o=null;
        setParams(sql, params);              //根据 sql 语句和 params,设置 pstmt 对象
        rs=pstmt.executeQuery();
        if(rs.next())
            o=rs.getObject(1);
        returno;
    }
```

```java
//将结果集中封装成一个List
private static List getListFromRS()throws NamingException, SQLException {
    List list=newArrayList();
    //获取元数据
    ResultSetMetaData rsmd=rs.getMetaData();
    while(rs.next()){
        Map m=newHashMap();
        for(inti=1; i<=rsmd.getColumnCount(); i++){
            //获取当前行第i列的数据类型
            String colType=rsmd.getColumnTypeName(i);
            //获取当前行第i列的列名
            String colName=rsmd.getColumnName(i);
            String s=rs.getString(colName);
            if(s !=null){
                System.out.println(colType+colName);
                if(colType.equals("INTEGER")||colType.equals("INT"))
                    m.put(colName, new Integer(rs.getInt(colName)));
                elseif(colType.equals("FLOAT"))
                    m.put(colName, new Float(rs.getFloat(colName)));
                else{
                    //其余类型均作为String对象取出
                    m.put(colName, rs.getString(colName));
                    //System.out.println("==="+m);
                }
            }
        }
        list.add(m);
    }
    returnlist;
}
//查询获取List对象
public static List getList(String sql, Object[] params){
    List list=null;                    //定义保存查询结果的集合对象
    try {
        setParams(sql, params);        //根据sql语句和params,设置pstmt对象
        rs=pstmt.executeQuery();       //执行SQL语句,得到结果集
        list=getListFromRS();          //根据RS得到list
    } catch(Exception e){
        e.printStackTrace();
    } finally {
        close();
    }
    returnlist;
}
```

```
        public static List getList(String sql){
            returngetList(sql, new Object[] {});
        }
        //查询获得 int 型数
        public staticint getInt(String sql, Object[] params){
            inti=0;
            try {
                Object temp=getObjectFromRS(sql, params);
                if(temp!=null)
                    i=((Integer)temp).intValue();
            } catch(Exception e){
                e.printStackTrace();
            } finally {
                close();
            }
            returni;
        }
        publicstaticint getInt(String sql){
            returngetInt(sql, new Object[] {});
        }
    }
```

12.5.3 项目说明

LoginActivity 中调用了封装网络通信操作的工具类 AsynHttp。创建 AsynHttp 对象时,需要给出三个参数,即连接的 URL、当前的 Context 对象、实现回调函数接口对象。其中,URL 即为服务端登录 Servlet 的 URL 地址,Context 对象使用当前 Activity 的上下文对象,实现回调函数接口对象同样为当前 Activity,即 LoginActivity。这样就可以在 LoginActivity 中实现回调接口方法 callBackFunction(),该方法用于异步任务后台线程的网络通信操作执行完后处理服务器端响应数据。

服务器端的登录 Servlet(LoginAction)获取客户端提交的请求参数后,调用 Service 方法,连接数据库,判断用户名和密码是否正确,并根据判断结果生成对应的 JSON 格式的响应数据,并输出到客户端。

接收到服务端响应后,客户端会自动调用回调函数 callBackFunction(),在其中解析返回的 JSON 数据,根据解析结果判断是否登录成功,若成功,则跳转到 MainActivity,否则提示出错信息。

12.6 项目实战:店铺列表功能

12.6.1 项目分析

在 CoffeeStore 的主页上单击"店铺",打开店铺列表界面,以列表形式显示当前所有店铺信息。每个店铺信息包括店铺图片、店铺名称、店铺地址、联系电话。店铺信息来自服务器端程序返回的数据,需要在异步任务中与服务器端程序进行通信,获取返回的数据并进行解析,将相应数据显示在 ListView 控件的每一个列表项上。

12.6.2 项目实现

首先创建 1 个 Activity,表示店铺列表页,Activity 名称如下。

● ShopActivity

在 ShopActivity 中使用 HttpURLConnection 进行与服务器端程序的网络通信,解析服务器端返回的 JSON 数据,并将数据显示在 ShopActivity 的 ListView 组件上。

ShopActivity 的代码如下。

```java
public class ShopActivity2 extends Activity implements IAsynHttpCallBack,
IAsynGetImageCallBack{
   ArrayList<HashMap<String,Object>>data;
   private  ListView list;
MyAdapter adapter;
@Override
protected void onCreate(Bundle savedInstanceState){
super.onCreate(savedInstanceState);
setContentView(R.layout.shop_layout);
list=(ListView)findViewById(R.id.listshop);
showAllShops();
}
@Override
protected void onRestart(){
//TODO Auto-generated method stub
super.onRestart();
showAllShops();
}
public  void showAllShops(){
    AsynHttp ao = new AsynHttp ( " http://10.0.2.2:8080/CoffeeStoreServer/
    shopList.action",this, this);
ao.execute();
data=new ArrayList<HashMap<String, Object>>();
adapter=new MyAdapter
    (ShopActivity2.this, data, R.layout.list_item_custom,
    new String[]{"shop_name","shop_address","shop_tel","shop_img"},
    new int[]{R.id.txtName,R.id.txtAddress,R.id.txtTel,R.id.img});
adapter.setViewBinder(new SimpleAdapter.ViewBinder(){
@Override
public boolean setViewValue(View view, Object data,
String textRepresentation){
```

```java
if(view instanceof ImageView && data instanceof Bitmap){
            ImageView iv=(ImageView)view;
iv.setImageBitmap((Bitmap)data);
            return true;
}
return false;
}
    });
list.setAdapter(adapter);
}
class MyAdapter extends SimpleAdapter{
public MyAdapter(Context context, List<? extends Map<String, ?>> data, int resource, String[] from, int[] to){
super(context, data, resource, from, to);
}
@Override
public View getView(int position, View convertView, ViewGroup parent){
        View result=super.getView(position, convertView, parent);
TextView txtTilte=(TextView)result.findViewById(R.id.txtName);
        if(position%2==1){
            result.setBackgroundColor(Color.GREEN);
txtTilte.setTextColor(Color.BLUE);
}
else{
            result.setBackgroundColor(Color.YELLOW);
txtTilte.setTextColor(Color.RED);
}
return result;
}
   }
@Override
public void getHttpImgCallBackFunction(Map<String, Object>result){
if(result!=null){
        Bitmap bm=(Bitmap)result.get("Bitmap");
Map<String,Object>m=(Map<String,Object>)result.get("data");
        if(m!=null){
            m.put("shop_img", bm);
adapter.notifyDataSetChanged();
}
        }
   }
@Override
public void callBackFunction(JSONObject jsonobj){
```

```java
        if(jsonobj!=null){
              parse(jsonobj);
    getBitMap(data);
    adapter.notifyDataSetChanged();
    }
        }
    public void parse(JSONObject jo){
    if(jo !=null){
    JSONArray jsonArray=jo.optJSONArray("shopList");
    if(jsonArray !=null){
    int length=jsonArray.length();
                if(0<length){
    for(int i=0; i<length; i++){
                   JSONObject jsonObj=jsonArray.optJSONObject(i);
                   if(jsonObj ==null)
    continue;
    HashMap<String, Object>m=new HashMap<String, Object>();
    m.put("shop_name", jsonObj.optString("shop_name"));
    m.put("shop_address", jsonObj.optString("shop_address"));
    m.put("shop_tel", jsonObj.optString("shop_tel"));
    m.put("shop_imgurl", jsonObj.optString("shop_imgurl"));
    data.add(m);
    }
                Log.v("data", data.toString());
    }
            }
    }
        }
    public void getBitMap(ArrayList<HashMap<String, Object>>data){
    if(data !=null){
    for(Map<String, Object>m : data){
            String url=(String)m.get("shop_imgurl");
    AsynGetImage agi=new AsynGetImage(url, this, this);
    agi.set_data(m);
    agi.getImage();
    }
        }
      }
    }
```

店铺列表功能中也涉及与服务器端获取店铺列表程序的交互，服务器端程序使用 Servlet 接收客户端请求，并调用 Service 层进行业务操作，连接后台数据库，查找店铺信息表，并使用 JSON 格式将查询结果返回给客户端。服务器端获取店铺列表信息的 Servlet 代码如下。

```java
@WebServlet("/shopList.action")
public class ShopActionextends HttpServlet{
    @Override
    protected void doGet (HttpServletRequest request, HttpServletResponse
    response)throws ServletException, IOException {
        //TODO Auto-generated method stub
        doPost(request,response);
    }
    @Override
    protected void doPost (HttpServletRequest request, HttpServletResponse
    response)throws ServletException, IOException {
        //TODO Auto-generated method stub
        PrintWriter out=response.getWriter();
        ShopService ps=newShopService();
        List<Map<String,String>>sl=ps.findShopList();
        JSONArray jsonpl=JSONArray.fromObject(sl);
        String result="{\"shopList\":"+jsonpl.toString()+"}";
        out.print(result);
        out.flush();
        out.close();
    }
}
```

服务器端获取店铺列表信息的 Service 代码如下。

```java
public class ShopService {
    public List<Map<String, String>>findShopList(){
        //TODO Auto-generated method stub
        String sql="select * from shop order by shop_id";
        return DBUtil.getList(sql);
    }
}
```

其中调用的工具类 DBUtil 代码与上一节相同，在此不再赘述。

12.6.3 项目说明

ShopActivity 中放置了 ListView 组件，用于显示店铺列表信息。ListView 的 Adapter 的数据源 data 来自服务端程序返回的 JSON 数据的解析结果。在与服务端程序通信结束后，接收到返回的响应数据，在 callBackFunction()方法中调用 parse()方法解析返回的响应数据，将解析结果存入 data。遍历 data，取出每个店铺图片的 URL 地址，再次利用异步任务获取图片，在回调方法 getHttpImgCallBackFunction()中将获取到的 Bitmap 对象也存入 data 中，利用 Adapter 的 notifyDataSetChanged()方法更新 ListView 控件。

在服务端 Servlet 程序中调用 Service 方法查询店铺表中的所有信息，将查询结果封装为 JSON 格式,返回给客户端。

本 章 小 结

  Android 中提供的异步任务是一种更加轻量级和简单的方式,来完成线程之间的通信。通常,在执行网络通信这种可能引发 ANR 异常的耗时操作时,会将这些操作放在异步任务的后台线程中(doInBackground()方法)进行,更新 UI 界面的操作则放在异步任务的运行于主线程的方法中(onPreExecute()、onProgressUpdate()、onPostExecute()方法)来进行,而后台线程方法和主线程方法间的通信,则依靠方法的各个参数和返回值来完成。

  另外,服务器端程序返回的数据格式,目前常用的是 JSON 这种轻量级的数据交换格式,它比 XML 更小、更快,更易解析,大大降低了之前对数据解析的编程开销。Android 中提供了现成工具类,对 JSON 格式进行表示和解析。例如,使用 JSONObject 类型表示 JSON 对象,使用 JSONArray 类型表示 JSON 数组,使用 optXXX()系列方法获取 JSON 对象中的各个属性值。

  Android 提供的网络通信接口,常见的有两种,但是从 Android 6.0 开始,谷歌已经移除了对 Apache Http Client 的支持。因此,建议编写网络通信程序时,尽量使用谷歌官方推荐的 HttpURLConnection 通信接口。HttpURLConnection 中提供了创建连接、设置连接属性、创建输入输出流对象等方法,利用这些方法,可以完成 Android 客户端程序与服务器端程序的通信。

本 章 习 题

  1. 说明在什么场合需要使用异步任务。
  2. 列举在异步任务中的子线程中执行的方法有哪些。在主线程中执行的方法有哪些。
  3. 编写一个 Android 程序,实现在 Activity 界面的 TextView 控件上显示当前计数值,计数值每隔 1 秒增 1。使用异步任务实现该功能。
  4. 列举 JSON 和 XML 的不同以及各自的优缺点。
  5. 列出 HttpURLConnection 的主要方法以及它们各自的作用。
  6. 使用 HttpURLConnection 完成 CoffeeStore 项目的用户注册功能。
  7. 假设有如下的 XML 数据,请给出对应的 JSON 数据格式。

```
<persons>
    <person name="小张" gender="男" age=26>
    <person name="小李" gender="女" age=24>
</persons>
------
{
```

```
"persons":[
{ "name":"小张","gender":"男","age":26},
{ "name":"小李","gender":"女","age":24}
]
}
```

8. 假设有一个 Activity,需要从服务器获取菜品的信息,并存储到本地数据中,完成以下功能。

(1) 设计出菜品所包含的信息。

(2) 给出合理的设计界面(可以直接绘出)。

(3) 设计出服务器提供的接口及参数(假定服务器为 http://192.168.0.1/DRSS/)。

(4) 设计出合理的 JSON 数据格式,便于从服务器获取数据。

(5) 设计出本地数据库中存储菜品信息的表的结构。

第4篇

Android 高级开发篇

深入学习 Intent

本章概述

本章讲解 Intent 过滤器的解析规则，PendingIntent 的定义与应用场景，以及通过 Intent 实现发短信、打电话的功能，运用 PendingIntent 实现系统通知栏的功能。

学习重点与难点

重点：

（1）PendingIntent 对象的应用。

（2）Intent 过滤器的解析规则。

难点：

（1）掌握使用 Intent 来收发短信、拨打电话的功能。

（2）通过 PendingIntent 实现系统通知栏的功能。

学习建议

学习本章之前，回顾第 7 章的 Intent 基本概念，有助于深入理解 Intent，在了解 Intent 基本概念的基础上更深入地学习 PendingIntent 对象，通过 PendingIntent 实现系统通知栏的效果。进而掌握 Intent 过滤器的解析机制，对 Intent 进行匹配与识别，最后，通过发信息、打电话等案例加深对 Intent 的整体应用与理解。

13.1 PendingIntent

　　Intent 表示一种意图，描述了想要启动一个 Activity、Broadcast 或 Service 的意图。它主要持有的信息是想要启动的组件（Activity、Broadcast 或 Service），在开发操作中，需要通过 startActivity、startService 或 sendBroadcast 方法来启动这个意图，执行某些操作。PendingIntent 可认为是对 Intent 的包装，PendingIntent 主要持有的信息是它所包装的 Intent 和当前应用程序 Context，即使当前应用程序已经不存在了，也能通过存在于 PendingIntent 里的 Context 来执行 Intent。当把 PendingIntent 递交给别的程序进行处理时，PendingIntent 仍然拥有 PendingIntent 原程序所拥有的权限，当从系统取得一个 PendingIntent 时，一定要确保 Intent 最终能发到目的组件，否则 Intent 可能不确定具体目标。

　　获取一个 PendingIntent 对象实例的方法如下。

（1）可通过 getActivity(Context context，int requestCode，Intent intent，int flags)方法从系统取得一个用于启动一个 Activity 的 PendingIntent 对象。

（2）可通过 getService(Context context，int requestCode，Intent intent，int flags)方法从系统取得一个用于启动一个 Service 的 PendingIntent 对象。

（3）可通过 getBroadcast(Context context，int requestCode，Intent intent，int flags)方法从系统取得一个用于向 BroadcastReceiver 发送广播的 PendingIntent 对象。

而获取到 PendingIntent 对象后，需要设置相应的参数，其参数常量如下。

（1）FLAG_CANCEL_CURRENT：如果当前系统中已经存在一个 PendingIntent，将会取消存在的 PendingIntent，从而创建一个新的 PendingIntent 对象。

（2）FLAG_UPDATE_CURRENT：如果 AlarmManager 管理的 PendingIntent 已经存在，可以让新的 Intent 更新之前 PendingIntent 中的 Intent 对象数据，例如，更新 Intent 中的 Extras。此外，也可以在 PendingIntent 的原进程中调用 PendingIntent 的 cancel()方法，将其从系统中清除。

（3）FLAG_NO_CREATE：如果 AlarmManager 管理的 PendingIntent 已经存在，那么将不进行任何操作，直接返回已经存在的 PendingIntent；如果 PendingIntent 不存在，返回值为空。

13.2 Intent 过滤器

在隐式启动 Activity 时，并没有在 Intent 中指明 Activity 所在的类。因此，Android 系统一定存在某种匹配机制，使 Android 系统能够根据 Intent 中的数据信息找到需要启动的 Activity。而这种匹配机制是依靠 Android 系统中的 Intent 过滤器(Intent-Filter)实现的。

Intent 过滤器是一种根据 Intent 中的动作(action)、类别(category)和数据(data)等信息，对适合接收 Intent 的组件进行匹配和筛选的机制。Intent 过滤器可以匹配数据类型、路径以及相关协议，也可以确定多个匹配项顺序的优先级别(priority)。应用程序的 Activity、Service、BroadcastReceiver 组件都能够注册 Intent 过滤器，因此，这些组件在相应的数据类型上则可以产生相应的动作。

如果要让组件能够注册 Intent 过滤器，需要在 AndroidManifest.xml 文件中的各个组件下定义＜intent-filter＞标签节点，再在节点内声明组件可支持的动作、类别、数据等信息。也可以在 Java 代码中动态地设置 Intent 过滤器的基本信息。＜intent-filter＞节点支持的标签和属性请参考表 13.1。

表 13.1 ＜intent-filter＞节点属性

标　　签	属　　性	描　　述
＜action＞	android:name	组件所响应的动作，用字符串描述，通常由 Java 类名和包名构成，如 cn.edu.neusoft.intent.MyFilter。
＜category＞	android:category	指定 Intent 请求的动作类别

续表

标签	属性	描述
<data>	android:host	指定一个有效的主机名
	android:minetype	指定组件所处理的数据类型
	android:path	有效的 URI 路径
	android:port	有效的端口号
	android:scheme	需要的模式

<category>标签属性用来指定 Intent 过滤器的匹配类别,每个 Intent 过滤器可定义多个<category>标签,开发人员可以定义自己的类别信息,也可以使用系统提供的类别。Android 提供的系统类别请参考表 13.2。

表 13.2 Android 提供的系统类别

类别名称	描述
ALTERNATIVE	Intent 数据默认动作的一个可替换执行方法
SELECTED_ALTERNATIVE	同 ALTERNATIVE 相似,而是被解析出来的执行方法
BROWSABLE	声明 Activity 可由浏览器启动
DEFAULT	为 Intent 过滤器所定义的数据提供默认动作
HOME	设备启动后显示的第一个 Activity
LAUNCHER	在应用程序启动时首先被显示

Intent 解析,即 Intent 到 Intent 过滤器的映射过程。Intent 解析可以在所有组件中找到一个与请求 Intent 达成完全匹配的 Intent 过滤器。在 Android 系统中,Intent 解析的匹配规则如下。

(1) Android 系统把所有应用程序包中的 Intent 过滤器集合在一起,形成一个完整的 Intent 过滤器列表。

(2) 在 Intent 与 Intent 过滤器匹配时,Android 系统会将列表中所有 Intent 过滤器的动作与类别与 Intent 携带信息进行匹配识别,任何不匹配的 Intent 过滤器都将被过滤掉。而没有指定动作的 Intent 过滤器可匹配任意的 Intent,但是没有指定类别的 Intent 过滤器只能匹配没有类别的 Intent。

(3) 把 Intent 数据 URI 信息与 Intent 过滤器的<data>标签中的属性进行匹配,如果<data>标签指定了 host、port、minetype、path 信息,就都需要与 Intent 的 URI 数据进行匹配,任何不匹配的 Intent 过滤器均被过滤掉。

(4) 如果 Intent 过滤器的匹配结果不止一个,则可根据<intent-filter>标签中定义的优先级别来对 Intent 过滤器进行排序,选择优先级别最高的 Intent 过滤器。

通过下面的代码整体地理解 intent-filter 的过滤机制,首先需要在 AndroidManifest.xml 文件中注册 Intent 过滤器,代码如下。

```xml
<?xml version="1.0" encoding="utf-8"?>
<manifest xmlns:android="http://schemas.android.com/apk/res/android"
    package="cn.edu.neusoft.intentfiltertest"
    android:versionCode="1"
    android:versionName="1.0">
    <uses-sdk android:minSdkVersion="23" />
    <application
        android:icon="@drawable/ic_launcher"
        android:label="@string/app_name">
        <activity
            android:name=".MainActivity"
            android:label="@string/app_name">
            <intent-filter>
                <action android:name="android.intent.action.MAIN" />
                <category android:name="android.intent.category.LAUNCHER" />
            </intent-filter>
        </activity>
        <activity
            android:name=".NewActivity"
            android:label="@string/app_name">
            <intent-filter>
                <action android:name="android.intent.action.VIEW" />
                <category android:name="android.intent.category.DEFAULT" />
                <data
                    android:host="edu.neusoft"
                    android:scheme="schemetest" />
            </intent-filter>
        </activity>
    </application>
</manifest>
```

在 AndroidManifest.xml 文件中定义的过滤器的动作是 android.intent.action.VIEW,表示根据 URI 协议,通过浏览的方式启动相应的 Activity,类别指明的是 android.intent.category.DEFAULT,表示默认动作,数据的 scheme 是 schemetest,主机部分是 edu.neusoft。

在 MainActivity 中定义一个 Intent 来启动 NewAcitivity,而这个 Intent 与 Activity 设置的 Intent 过滤器是完全匹配的,相关代码如下。

```
Intent intent = new Intent(Intent.ACTION_VIEW, Uri.parse("schemetest://edu.neusoft/path"));
startActivity(intent);
```

Intent 与 AndroidManifest.xml 中定义的 Intent 过滤条件完全一致,进行匹配时,完全符合所定义的条件,那么 Intent 与 Intent 过滤器的匹配结果成功,则从 MainActivity

跳转到 NewActivity。

13.3 运行时权限

2015 年 8 月,谷歌发布了 Android 6.0 版本,代号叫做"棉花糖"(Marshmallow),其中很大的一部分变化是在用户权限的授权机制上,因为低于 6.0 版本的默认授权不合理。在 Android 6.0 版本中,权限授权更符合用户的操作习惯,只有在用户需要使用权限时才去授权请求,这样可以提高用户体验性,保障系统的安全度。

在 Android 6.0 中,权限分为两类:系统权限(SYSTEM PERMISSIONS)和特殊权限授权(SPECIAL GRANT)。在开发的过程中,系统权限分为两类,一类是 Normal Permissions(正常权限),此类权限一般不涉及隐私,不需要用户进行授权,比如震动、访问网络等;另一类是 Dangerous Permissions(危险权限),涉及隐私,需要用户手动授权,比如读取 sdcard、访问通讯录等,此类将每个单独权限进行分组。以下为分类下的所有权限。

Normal Permissions(正常权限):
- ACCESS_LOCATION_EXTRA_COMMANDS
- ACCESS_NETWORK_STATE
- ACCESS_NOTIFICATION_POLICY
- ACCESS_WIFI_STATE
- BLUETOOTH
- BLUETOOTH_ADMIN
- BROADCAST_STICKY
- CHANGE_NETWORK_STATE
- CHANGE_WIFI_MULTICAST_STATE
- CHANGE_WIFI_STATE
- DISABLE_KEYGUARD
- EXPAND_STATUS_BAR
- GET_PACKAGE_SIZE
- INSTALL_SHORTCUT
- INTERNET
- KILL_BACKGROUND_PROCESSES
- MODIFY_AUDIO_SETTINGS
- NFC
- READ_SYNC_SETTINGS
- READ_SYNC_STATS
- RECEIVE_BOOT_COMPLETED
- REORDER_TASKS
- REQUEST_INSTALL_PACKAGES

- SET_ALARM
- SET_TIME_ZONE
- SET_WALLPAPER
- SET_WALLPAPER_HINTS
- TRANSMIT_IR
- UNINSTALL_SHORTCUT
- USE_FINGERPRINT
- VIBRATE
- WAKE_LOCK
- WRITE_SYNC_SETTINGS

Dangerous Permissions(危险权限)：

group：android. permission-group. CONTACTS
- permission：android. permission. WRITE_CONTACTS
- permission：android. permission. GET_ACCOUNTS
- permission：android. permission. READ_CONTACTS

group：android. permission-group. PHONE
- permission：android. permission. READ_CALL_LOG
- permission：android. permission. READ_PHONE_STATE
- permission：android. permission. CALL_PHONE
- permission：android. permission. WRITE_CALL_LOG
- permission：android. permission. USE_SIP
- permission：android. permission. PROCESS_OUTGOING_CALLS
- permission：com. android. voicemail. permission. ADD_VOICEMAIL

group：android. permission-group. CALENDAR
- permission：android. permission. READ_CALENDAR
- permission：android. permission. WRITE_CALENDAR

group：android. permission-group. CAMERA
- permission：android. permission. CAMERA

group：android. permission-group. SENSORS
- permission：android. permission. BODY_SENSORS

group：android. permission-group. LOCATION
- permission：android. permission. ACCESS_FINE_LOCATION
- permission：android. permission. ACCESS_COARSE_LOCATION

group：android. permission-group. STORAGE
- permission：android. permission. READ_EXTERNAL_STORAGE
- permission：android. permission. WRITE_EXTERNAL_STORAGE

group：android. permission-group. MICROPHONE
- permission：android. permission. RECORD_AUDIO

group:android.permission-group.SMS

- permission:android.permission.READ_SMS
- permission:android.permission.RECEIVE_WAP_PUSH
- permission:android.permission.RECEIVE_MMS
- permission:android.permission.RECEIVE_SMS
- permission:android.permission.SEND_SMS

permission:android.permission.READ_CELL_BROADCASTS

对于 Dangerous Permissions，每一个权限是以分组形式区分的。权限分组机制对于开发 App 也存在一定影响，当运行在 Android 6.0 环境时，如果 App 已被授权了同一组的某个权限，再申请某个权限时，系统会立即进行授权，不需要手动单击授权操作。例如，程序已经对 READ_CONTACTS 进行授权，如果再申请同组内的 WRITE_CONTACTS 权限，系统会直接通过授权，弹出的对话框将提示整个权限组的说明，而不是单个权限。所以，此版本的权限分组可能有一些不太友好，后续的版本会进一步改进。

了解了 Android6.0 下的权限授予机制后，在开发过程中，如何对指定权限进行控制，以保证系统运行的安全性呢？以下案例讲解与实现权限 API，以便应用在项目开发中。

【例 13-1】 本案例演示运行时权限的处理，以发送短信权限的授权控制为例，操作演示申请权限（发送短信）以及处理权限回调。

(1) 在 AndroidManifest.xml 文件中添加需要的权限。

此步骤同前版本一样，都需要在 AndroidManifest.xml 文件中添加相应的权限。代码如下。

```
<uses-permission android:name="android.permission.SEND_SMS" />
```

(2) 检查权限。对要使用的权限进行检测，是否已被授权。代码如下。

```
ContextCompat.checkSelfPermission(SmsSendActivity.this, Manifest.permission.SEND_SMS);
```

返回值为

PackageManager.PERMISSION_DENIED、PackageManager.PERMISSION_GRANTED。

当返回 PERMISSION_DENIED，就需要进行申请授权。

(3) 请求权限进行授权。此方法是异步的，第一个参数是 Context；第二个参数是需要申请的权限的字符串数组；第三个参数为 requestCode，用于回调判断。同时也支持一次性申请多个权限，系统通过对话框逐一询问用户是否授权。代码如下。

```
ActivityCompat.requestPermissions(SmsSendActivity.this,
    new String[]{Manifest.permission.SEND_SMS},
        REQUEST_CODE_ASK_PERMISSIONS);
```

(4) 对申请权的回调处理。代码如下。

```
@Override
```

```java
public void onRequestPermissionsResult(int requestCode, String[] permissions,
int[] grantResults)
    {
    switch(requestCode){
        case REQUEST_CODE_ASK_PERMISSIONS:
            if(grantResults[0] ==PackageManager.PERMISSION_GRANTED){
                //Permission Granted
                sendSms();
            } else {
                //Permission Denied
                Toast.makeText(SmsSendActivity.this, "SEND_SMS Denied", Toast.
                LENGTH_SHORT).show();
            }
            break;
        default:
            super.onRequestPermissionsResult(requestCode, permissions,
            grantResults);
    }
}
```

在 REQUEST_CODE 判断权限授权成功后,处理相关业务逻辑,如发送短信等。发短信的方法 sendSms() 如下所示。

```java
public void sendSms(){
    String phoneNumber=((EditText)findViewById(R.id.number)).getText().
    toString();
    String smsContent=((EditText)findViewById(R.id.mainText)).getText().
    toString();

    SmsManager sms=SmsManager.getDefault();
    PendingIntent pi = PendingIntent. getActivity (this, 0, new Intent (this,
    IntentSendMessage.class), 0);
    if(smsContent.length()>70){
        ArrayList<String>msgs=sms.divideMessage(smsContent);
        for(String msg : msgs){
            sms.sendTextMessage(phoneNumber, null, msg, pi, null);
        }
    } else {
        sms.sendTextMessage(phoneNumber, null, smsContent, pi, null);
    }
    Toast.makeText(SmsSendActivity.this, "短信发送完成", Toast.LENGTH_SHORT).
    show();
}
```

本案例完整演示了 Android 6.0 下权限的处理机制,之后的案例将不再演示权限相关问题,请参照本程序的具体实现完善其他程序。

程序运行效果如下。

运行首页,如图 13.1 所示,单击"使用授权处理"按钮,进入发送短信界面,如图 13.2 所示,输入相关信息后单击"发送"按钮,将会提示"需要授权发送信息权限",如图 13.3 所示;单击"同意授权"按钮,进入系统默认提示对话框,如图 13.4 所示。如果拒绝,授权失败,如图 13.5 所示;如果允许,则授权成功,短信发送完成,如图 13.6 所示。

图 13.1　首页

图 13.2　发短信界面

图 13.3　授权提示界面

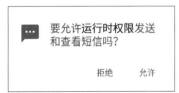

图 13.4　系统默认提示对话框

如果单击"未用授权处理"按钮,则省略授权提示过程,直接发送短信成功。如图 13.6 所示。

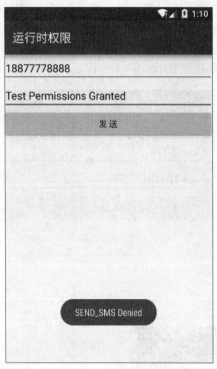

图 13.5 禁止授权提示

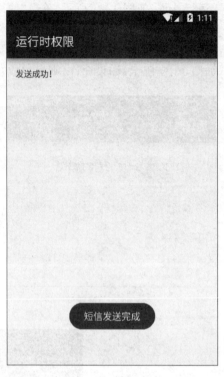

图 13.6 授权成功

【例 13-2】 发送短消息。

实现发送短消息功能,需要定义显示收件人、信息内容的控件,单击发送短消息按钮,将短信发送出去,并提示短消息发送成功。

布局文件定义的代码如下。

```
<?xml version="1.0" encoding="utf-8"?>
<LinearLayout xmlns:android="http://schemas.android.com/apk/res/android"
    android:layout_width="fill_parent"
    android:layout_height="fill_parent"
    android:background="#FFFFFF"
    android:orientation="vertical">
    <LinearLayout
        android:layout_width="fill_parent"
        android:layout_height="wrap_content"
        android:layout_marginTop="10px"
        android:orientation="horizontal">
        <TextView
            android:id="@+id/TextView01"
            android:layout_width="wrap_content"
```

```xml
            android:layout_height="wrap_content"
            android:text="收件人:"></TextView>
        <EditText
            android:id="@+id/addressee"
            android:layout_width="fill_parent"
            android:layout_height="wrap_content"></EditText>
    </LinearLayout>
    <EditText
        android:id="@+id/message"
        android:layout_width="fill_parent"
        android:layout_height="200px"
        android:gravity="top"></EditText>
    <Button
        android:id="@+id/send"
        android:layout_width="wrap_content"
        android:layout_height="wrap_content"
        android:layout_gravity="center"
        android:text="发送信息"></Button>
</LinearLayout>
```

在AndroidManifest.XML文件中注册发送短消息的权限。代码如下。

```xml
<uses-permission android:name="android.permission.SEND_SMS" >
</uses-permission>
```

实现短消息发送的核心需要构建一个SmsManager，通过sendTextMessage()方法设置收件人、短信内容等信息，单击"发送消息"按钮，会收到相应短消息发送成功的提示，Activity中的代码实现如下。

```java
package com.sendsms;

importandroid.app.Activity;
import android.app.AlertDialog;
import android.app.PendingIntent;
import android.content.DialogInterface;
import android.content.Intent;
import android.os.Bundle;
import android.telephony.SmsManager;
import android.view.View;
import android.view.Window;
import android.widget.Button;
import android.widget.EditText;
import android.widget.Toast;
public class SendActivity extends Activity {
    private Button sendButton=null;          //创建发送按钮Button组件对象
    private EditText addressee=null;         //创建收件人编辑框EditText组件对象
```

```java
private EditText message=null;                    //创建信息内容编辑框 EditText 组件对象
@Override
public void onCreate(Bundle savedInstanceState){
super.onCreate(savedInstanceState);
requestWindowFeature(Window.FEATURE_NO_TITLE);
setContentView(R.layout.main);
sendButton= (Button)findViewById(R.id.send);
                                                  //实例化发送按钮 Button 组件对象
addressee= (EditText)findViewById(R.id.addressee);
                                                  //实例化收件人编辑框 EditText 组件对象
message= (EditText)findViewById(R.id.message);
                                                  //实例化收件人编辑框 EditText 组件对象
addressee.setText("请输入接收人的电话号码");      //设置默认收件人提示信息
message.setText("请输入短信内容");                //设置默认信息内容提示信息
        //添加收件人编辑框单击事件监听
addressee.setOnClickListener(new EditText.OnClickListener(){
@Override
public void onClick(View arg0){
addressee.setText("");
}
        });
//添加信息内容编辑框单击事件监听
message.setOnClickListener(new EditText.OnClickListener(){
@Override
public void onClick(View arg0){
   message.setText("");
}
        });
//添加发送按钮单击事件监听
sendButton.setOnClickListener(new Button.OnClickListener(){
@Override
public void onClick(View arg0){
             String strAddressee=addressee.getText().toString();
                                                  //获取收件人信息
String strMessage=message.getText().toString();
                                                  //获取发送内容消息
if("".equals(strAddressee)){                      //判断收件人信息是否为空
showMessage("收件人信息不能为空");                //调用信息提示方法
return;
}
if("".equals(strMessage)){                        //判断发送内容是否为空
showMessage("信息内容不能为空");                  //调用信息提示方法
return;
}
```

```
SmsManager smsManager=SmsManager.getDefault();
//构建 PendingIntent 对象,并使用 getBroadcast()广播
PendingIntent pendingIntent=PendingIntent.getBroadcast(
                SendActivity.this, 0, new Intent(), 0);
smsManager.sendTextMessage(strAddressee, null, strMessage,
pendingIntent, null);                    //发送短信消息
Toast.makeText(SendActivity.this, "短信发送成功", 1000).show();
                                        //信息提示方法
}
    });
}
public void showMessage(String message){
        AlertDialog alertDialog=new AlertDialog.Builder(this).create();
                                //创建 AlertDialog 对象
alertDialog.setTitle("提示信息");       //设置信息标题
alertDialog.setMessage(message);       //设置信息内容
        //设置确定按钮,并添加按钮监听事件
alertDialog.setButton("确定",
            new android.content.DialogInterface.OnClickListener(){
@Override
public void onClick(DialogInterface arg0, int arg1){
            }
        });
alertDialog.show();                    //设置弹出提示框
}
}
```

启动两个模拟器,一个模拟器用来发送短消息,另一个用来接收短消息,填写收件人与短信内容,单击"发送信息"按钮,用于接收短信的模拟器便会收到刚刚发送的信息,如图 13.7 所示。

【例 13-3】 拨打电话。

本程序建立在深入掌握 Intent 的基础上,运用 Intent 的 action、data 属性实现直接拨打电话以及启动拨号盘的功能。在布局文件中定义输入电话号码的可编辑文本控件(EditText),添加两个按钮控件,分别为直接拨打电话与拨打电话界面。

```
<?xml version="1.0" encoding="utf-8"?>
<LinearLayout xmlns:android="http://schemas.android.com/apk/res/android"
    android:layout_width="fill_parent" android:layout_height="fill_parent"
    android:background="#FFFFFF" android:orientation="vertical">
    <EditText android:id="@+id/phone_number"
        android:layout_width="fill_parent"
        android:layout_height="wrap_content"
        android:layout_gravity="center"
        android:hint="请输入号码">
```

(a)   (b)

图 13.7　发短信运行效果

```
    </EditText>
    <LinearLayout android:layout_width="wrap_content"
        android:layout_height="wrap_content"
        android:layout_gravity="center"
        android:background="#FFFFFF"
        android:orientation="horizontal">
        <Button android:id="@+id/phone_call"
            android:layout_width="wrap_content"
            android:layout_height="wrap_content"
            android:text="直接拨打电话"></Button>
        <Button android:id="@+id/phone_dial"
            android:layout_width="wrap_content"
            android:layout_height="wrap_content"
            android:text="拨打电话界面">
        </Button>
    </LinearLayout>
</LinearLayout>
```

在 MainActivity 类中定义按钮的单击监听器，分别创建 call() 与 dial() 两个方法实现直接拨打电话与拨打电话页面按钮的功能，通过获取输入的电话号单击按钮执行相应的操作。

```java
public class MainActivity extends Activity {
    private EditText editText=null;           //电话号码 EditText 组件对象
    private Button callButton=null;           //直接拨打按钮 Button 组件对象
    private Button dialButton=null;           //启动拨打界面按钮 Button 组件对象
    @Override
    public void onCreate(Bundle savedInstanceState){
        super.onCreate(savedInstanceState);
        setContentView(R.layout.main);
        editText=(EditText)findViewById(R.id.phone_number);
                                              //获取 EditText 控件对象
        CallActivity.this.registerForContextMenu(editText);
        callButton=(Button)findViewById(R.id.phone_call);
                                              //获取拨打按钮控件对象
        dialButton=(Button)findViewById(R.id.phone_dial);
                                              //获取拨打界面按钮控件对象
        //添加 Button 按钮单击监听
        callButton.setOnClickListener(new Button.OnClickListener(){
            @Override
            public void onClick(View arg0){
                call();                        //调用直接打电话的方法
            }
        });
        //添加 Button 按钮单击监听
        dialButton.setOnClickListener(new Button.OnClickListener(){
            @Override
            public void onClick(View arg0){
                dial();                        //调用启动一个拨号器的方法
            }
        });
    }
    /**
     * 直接打电话的方法
     */
    public void call(){
        String data="tel:"+editText.getText(); //电话号码参数字符串
        Uri uri=Uri.parse(data);               //将字符串转化为 Uri 实例
        Intent intent=new Intent();            //实例化 Intent
        intent.setAction(Intent.ACTION_CALL);  //设置 Intent 的 Action 属性
        intent.setData(uri);                   //设置 Intent 的 Data 属性
        startActivity(intent);                 //启动 Activity
    }
```

```
/**
 * 启动一个拨号器的方法
 */
public void dial(){
    String data="tel:"+editText.getText();    //电话号码参数字符串
    Uri uri=Uri.parse(data);                  //将字符串转化为 Uri 实例
    Intent intent=new Intent();               //实例化 Intent
    intent.setAction(Intent.ACTION_DIAL);     //设置 Intent 的 Action 属性
    intent.setData(uri);                      //设置 Intent 的 Data 属性
    startActivity(intent);                    //启动 Activity
}
```

执行程序,单击"直接拨打电话"按钮,或单击"拨打电话界面"按钮,将会跳到拨号盘界面,再按拨号键拨打电话,运行效果如图 13.8 所示。

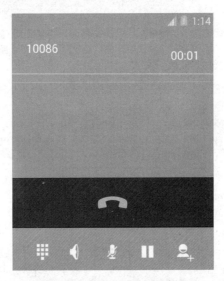

图 13.8 拨打电话界面

【例 13-4】 系统通知栏。

本程序通过 PendingIntent 对象实现发送系统通知消息功能。在布局文件中添加两个按钮,用于发送通知以及清除通知栏中的通知,具体代码如下。

```
<LinearLayout xmlns:android="http://schemas.android.com/apk/res/android"
    xmlns:tools="http://schemas.android.com/tools"
    android:layout_width="fill_parent"
    android:layout_height="wrap_content"
    android:orientation="vertical"
    tools:context="cn.edu.neusoft.notification.MainActivity">
    <Button
        android:id="@+id/start"
```

```xml
        android:layout_width="fill_parent"
        android:layout_height="wrap_content"
        android:text="发送通知" />
    <Button
        android:id="@+id/cancel"
        android:layout_width="fill_parent"
        android:layout_height="wrap_content"
        android:text="清除通知" />
</LinearLayout>
```

在 MainActivity 类中首先获取到系统的通知管理器（NotificationManager）服务，定义 Notification 对象、PendingIntent 对象，如用真机测试，设置震动效果，需在 Manifest 文件中添加震动权限许可，具体实现代码如下。

```java
public class MainActivity extends Activity {
    private Button start;
    private Button cancel;
    private static final int ID=1;
    private NotificationManager iNotificationManager;
    private Notification iNotification;

    @Override
    public void onCreate(Bundle savedInstanceState){
        super.onCreate(savedInstanceState);
        setContentView(R.layout.main);

        start= (Button)findViewById(R.id.start);
        cancel= (Button)findViewById(R.id.cancel);
        //1 获取 NotificationManager
        iNotificationManager=  ( NotificationManager )  getSystemService
        (NOTIFICATION_SERVICE);
        //2 初始化 Notification
        //API 版本在 23 以下可通过如下方式实例化 Notification 对象
        //int icon=R.drawable.nba;
        //CharSequence text="NBA 季后赛最新消息";
        //long time=System.currentTimeMillis();
        //iNotification=new Notification(icon, text, time);

        //iNotification.defaults=Notification.DEFAULT_ALL;
        //iNotification.flags=Notification.FLAG_NO_CLEAR;
        //iNotification.flags=Notification.FLAG_SHOW_LIGHTS;
        //设置声音
        //iNotification.sound=Uri.parse("file:///sdcard/msg.mp3");
        //设置振动
        //iNotification.vibrate=new long[]{0, 50, 100, 150};
```

```
//设置闪光灯颜色
iNotification.ledARGB=R.color.RED;
//设置闪光灯多久熄灭
iNotification.ledOffMS=800;
//设置闪光灯多久开启
iNotification.ledOnMS=800;
//3 定义 Notification 的消息、PendingIntent
CharSequence title="NBA 新闻,勇士迎来创队史的 23 连胜纪录";
CharSequence content="休息了两天的金州勇士队等来了今日的对手印第安纳步行
者队,后者通过赢得了西部 8 强资格";
PendingIntent contentIntent=PendingIntent.getActivity(
        MainActivity.this, 0, super.getIntent(),
        PendingIntent.FLAG_UPDATE_CURRENT);
iNotification.setLatestEventInfo(MainActivity.this, title, content,
        contentIntent);
//下面代码为 API-23 以上版本的实现方式
final Notification.Builder builder=new Notification.Builder
(MainActivity.this)
        .setSmallIcon(R.drawable.nba)
        .setContentTitle(title)
        .setContentText(content)
        .setContentIntent(contentIntent);
iNotification=builder.build();
//4 设置 NotificationManager 信息的发送及取消
//发送 Notification
start.setOnClickListener(new OnClickListener(){
    @Override
    public void onClick(View v){
        iNotificationManager.notify(ID, iNotification);
    }
});
//消除 Notification
cancel.setOnClickListener(new OnClickListener(){

    @Override
    public void onClick(View v){
        iNotificationManager.cancel(ID);
    }
});
    }
}
```

执行程序的页面显示如图 13.9 所示。

单击"发送通知"按钮,在通知栏中有通知消息提示,下拉通知栏将会查看到通知,包括图标、标题、内容等相关信息,单击"清除通知"按钮,将删除通知栏中的通知,显示效果如图 13.10 所示。

图 13.9 程序运行结果

图 13.10 通知栏界面

本 章 小 结

本章主要涉及 Android 体系中 Intent 过滤器的思想。然后运用示例讲解 Intent 过滤机制,实现拨打电话、接发短信等基本实用的功能,以及用 PendingIntent 实现系统通知栏的相关特性。

本 章 习 题

1. 简述 PendingIntent 的概念和应用场景。
2. 简述 IntentFilter 的解析规则。
3. 系统通知栏(Notification)的实现涉及了哪些知识与内容?
4. 通过系统通知,如何理解消息推送的实现思路?请调研极光推送、百度、腾讯等第三方推送方案的具体实现方式。

项目实践:为 CoffeeStore 项目增加最新上线商品推送功能,可使用第三方推送实现。

广播与服务

本章概述

本章讲解广播的概念、广播的实现方式,服务(Service)的基本概念、服务的隐式启动与显式启动、服务的生命周期。

学习重点与难点

重点:

(1) 广播的基本概念。
(2) 广播接收器的实现思想。
(3) 服务的概念和用途。
(4) 服务的生命周期。
(5) 服务的启动方式。
(6) 服务与 Broadcast 的综合案例运用。

难点:

(1) 掌握实现截获短信、显示来电位置的思想。
(2) 实现异步接收广播消息的方法。
(3) 理解服务生命周期的思想。
(4) 服务与 Broadcast 的综合案例运用。

学习建议

学习本章的知识前,可搜索相关资料,理解广播机制的思想,再通过广播在手机移动端的应用实例更加深入理解广播接收器的具体逻辑思想。最后,以实际案例加深掌握广播接收器的概念,分析日常生活中的应用场景。

理解 Service 的基本思想后,掌握好服务执行的生命周期情况,分析音乐播放器、上传/下载文件的实现原理,运用服务开发音乐播放器等相关的应用程序。

14.1 广播的定义与用途

在 Android 中,广播是一种广泛运用在应用程序之间传输信息的机制。广播分为两个方面:广播发送者和广播接收者。通常情况下,Broadcast Receiver 指的就是广播接收者(广播接收器)。Broadcast Receiver 是对发送出来的广播消息进行过滤接收并响应的一类组件,比如电池的使用状态、电话的接收和短信的接收都会产生一个广播,应用程序开发人员也可以监听这些广播,并做出相应的逻辑处理。

从实现原理上看,Android 中的广播使用了观察者模式,基于消息的发布/订阅事件模型。因此,从实现的角度来看,Android 中的广播机制将广播的发送者和接受者在极大程度上解耦,而使系统方便集成,更易扩展。

广播作为 Android 组件间的通信方式,可应用的场景如下。
- App 内部同一组件内的消息通信(单个或多个线程之间)。
- App 内部不同组件之间的消息通信(单个进程)。
- App 多个进程不同组件之间的消息通信。
- 不同 App 之间组件的消息通信。
- Android 系统在特定情况下与 App 之间的消息通信。

14.2　广播接收器的实现

广播机制的实现与 Intent 是密切相关的,而 Intent 的另一种用途便是发送广播消息。广播消息的内容可以是与程序相关的数据信息,也可以是 Android 系统信息,包括网络连接变化、电池电量变化、短信接收情况等。如果应用程序注册了 Broadcast Receiver,则可以接收到相应的广播消息。实现广播接收器的流程如下。

(1) 广播接收者 Broadcast Receiver 通过 Binder 机制向 AMS(Activity Manager Service)进行注册。

(2) 广播发送者通过 Binder 机制向 AMS 发送广播。

(3) AMS 查找符合相应条件(根据 Intent Filter/Permission 所设置条件)的 Broadcast Receiver,将广播发送到 Broadcast Receiver(一般情况下指 Activity)相应的消息循环队列中。

(4) 消息循环执行接收到此的广播消息,回调 Broadcast Receiver 中的 on Receive()方法。

实现广播接收器相对简单,只须创建一个 Intent,调用 sendBroadcast()函数就可把 Intent 携带的信息发送出去。代码如下。

```
Intent intent=new Intent();
intent.putExtra("msg", "This is Broadcast Message");
intent.setAction("cn.edu.neusoft.broadcastreceiver.SMS");
sendBroadcast(intent);
```

BroadcastReceiver 用于监听广播消息,在 AndroidManifest.XML 文件注册 BroadcastReceiver,通过 Intent 过滤器处理广播消息。代码如下。

```
<receiver android:name=".MyBroadcastReceiver" >
<intent-filter>
    <action android:name="cn.edu.neusoft.broadcastreceiver.SMS" />
</intent-filter>
```

```
</receiver>
```

创建 BroadcastReceiver 类接收发送的广播消息，并重写 onReceive()方法。代码如下。

```
public class MyBroadcastReceiver extends BroadcastReceiver {
    @Override
    public void onReceive(Context context, Intent intent){
        String msg=intent.getExtras().getString("msg");
        Toast.makeText(context, msg, 1000).show();
        System.out.println("Call onReceive()Method:Test BroadcastReceiver");
    }
}
```

当系统接收到与注册 BroadcastReceiver 匹配的广播消息时，就可以自动调用 BroadcastReceiver 的 onReceive()方法接收广播消息。

【例 14-1】 监视电池电量。

本程序主要运用广播接收器原理来实现对手机电池电量的监测，通过 Android 系统 Intent 常量 ACTION_BATTERY_CHANGED 获取电量的改变情况，在布局文件里定义显示电量变化的 TextView 控件，main.xml 代码如下。

```
<?xml version="1.0" encoding="utf-8"?>
<LinearLayout xmlns:android="http://schemas.android.com/apk/res/android"
    android:layout_width="fill_parent" android:layout_height="fill_parent"
    android:orientation="vertical">
    <TextView android:id="@+id/tvBatteryChanged" android:layout_width="fill_parent"
        android:layout_height="wrap_content" android:textSize="30dp" />
</LinearLayout>
```

在 MainActivity 类中定义一个 BroadcastReceiver 实例，重写 onReceive()方法，然后在 onCreate()方法里注册已定义的 BroadcastReceiver，代码如下。

```
public class Main extends Activity {
    private TextView tvBatteryChanged;
    private BroadcastReceiver batteryChangedReceiver=new BroadcastReceiver(){
        @Override
        public void onReceive(Context context, Intent intent){
            if(Intent.ACTION_BATTERY_CHANGED.equals(intent.getAction())){
                int level=intent.getIntExtra("level", 0);
                int scale=intent.getIntExtra("scale", 100);
                tvBatteryChanged.setText("电池用量:"+ (level * 100 / scale)+"%");
            }
        }
    };
    @Override
```

```
public void onCreate(Bundle savedInstanceState){
    super.onCreate(savedInstanceState);
    setContentView(R.layout.main);
    tvBatteryChanged=(TextView)findViewById(R.id.tvBatteryChanged);
    registerReceiver(batteryChangedReceiver, new IntentFilter(
        Intent.ACTION_BATTERY_CHANGED));
    }
}
```

执行程序运行结果如图 14.1 所示(建议用真机运行,监视电池电量的变化)。

图 14.1　监视电池电量效果图

【例 14-2】　发送广播消息。

本程序通过 BroadcastReceiver 发送广播消息,实现注册广播接收器、接收广播消息的过程。在布局文件中添加一个按钮、一个可编辑文本控件,分别为发送广播、输入广播的内容,完整代码如下。

```
<?xml version="1.0" encoding="utf-8"?>
<LinearLayout xmlns:android="http://schemas.android.com/apk/res/android"
    android:layout_width="fill_parent"
    android:layout_height="fill_parent"
    android:gravity="center_horizontal"
    android:orientation="vertical">
    <EditText
        android:id="@+id/notice"
        android:layout_width="fill_parent"
        android:layout_height="wrap_content"
        android:hint="请输入要发送的广播内容" />
    <Button
        android:id="@+id/send"
        android:layout_width="wrap_content"
        android:layout_height="wrap_content"
        android:text="发送广播" />
</LinearLayout>
```

新建类 MyBroadcastReceiver 继承 BroadcastReceiver 父类,重写 onReceive()方法,

用于接收发送的广播消息,完整代码如下。

```java
public class MyBroadcastReceiver extends BroadcastReceiver {
    @Override
    public void onReceive(Context context, Intent intent){
        String msg=intent.getExtras().getString("msg");
        Toast.makeText(
            context,
            "接收到广播消息"+"\n"+"消息内容:"+msg+"\n"+"action:"
                +intent.getAction(), 1000).show();
        System.out.println("Call onReceive()Method:Test BroadcastReceiver");
    }
}
```

在 MainActivity 类中,通过 sendBroadcast()方法实现发送广播消息,利用 Intent 过滤机制匹配 action 属性值,与 AndroidManifest.xml 文件中 action 完全一致,则广播发送成功。AndroidManifest.xml 的代码如下。

```xml
<?xml version="1.0" encoding="utf-8"?>
<manifest xmlns:android="http://schemas.android.com/apk/res/android"
    package="cn.edu.neusoft.broadcastreceiver"
    android:versionCode="1"
    android:versionName="1.0">
    <uses-sdk
        android:minSdkVersion="14"
        android:targetSdkVersion="17" />
    <application
        android:allowBackup="true"
        android:icon="@drawable/ic_launcher"
        android:label="@string/app_name"
        android:theme="@style/AppTheme">
        <activity
            android:name=".MainActivity"
            android:label="@string/app_name">
            <intent-filter>
                <action android:name="android.intent.action.MAIN" />
                <category android:name="android.intent.category.LAUNCHER" />
            </intent-filter>
        </activity>
        <receiver android:name=".MyBroadcastReceiver">
            <intent-filter>
                <action android:name="cn.edu.neusoft.broadcastreceiver.SMS" />
            </intent-filter>
        </receiver>
    </application>
```

```
</manifest>
```

配置好 AndroidManifest 文件，输入所要发送的广播消息内容，单击"发送广播"按钮，即成功发送了广播消息，浮动窗会显示广播的内容及 action 属性值，执行程序的运行效果如图 14.2 所示。

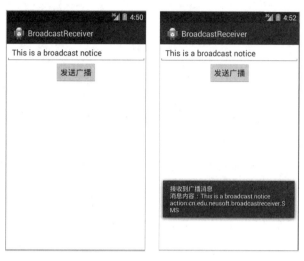

图 14.2　程序运行结果图

14.3　服务的基本概念

因为手机的硬件及屏幕大小的限制，通常情况下，只允许一个应用程序完整地显示在手机屏幕上，而暂停其他处于未激活状态的程序，而这样又不能充分利用手机。所以 Android 采用了一种服务机制，即允许在没有用户界面的情况下，使程序能够长时间在后台运行，提供应用程序的后台服务功能，并能够处理事件或数据更新的操作。

作为 Android 四大组件之一，服务（Service）的特性为不直接与用户交互，并长期在后台运行。而在实际的应用中，有很多情景需要应用服务，比如播放音乐、上传下载文件等。当下载文件或播放音乐时，不需要将程序一直运行并显示在手机屏幕上，可以在后台执行程序，如文件下载成功，可以发送提示消息，进而查看下载的内容。这样既充分利用了手机执行多任务，也提高了手机的使用效率。而服务除了实现后台服务功能，还可以用于进程间通信（Inter Process Communication），解决不同 Android 应用程序进程之间的调用和通信问题。

服务的使用方式有两种：一种是启动方式，另一种是绑定方式。

在启动方式中，通过调用 startService() 方法启动 Service，停止服务则调用 stopService() 或 stopSelf() 方法。因此，服务是由其他组件启动的，而停止可以通过其他组件或自身完成。以启动方式实现服务，启动服务的组件不能获取服务的对象实例，进而无法调用其中的任何函数，也无法获取其中的任何状态和数据信息。

启动服务的方法有两种：显示启动和隐式启动。显示启动需要在 Intent 中指明服务所在的类，调用 startService(Intent) 启动服务；而隐式启动需要在注册服务时声明 Intent-filter 的 action 属性。这样就可以在 Intent 的过滤机制下识别并匹配 action 属性，在不声明服务所在类的情况下启动服务。

隐式启动需要在注册服务时，声明 Intent-filter 的 action 属性。代码如下。

```
<service android:name=".RandomService">
<intent-filter>
    <action android:name="cn.edu.neusoft.service.RandomService" />
</intent-filter>
</service>
```

在 Activity 中启动服务时，需要设置 Intent 的 action 属性，在不声明服务所在类的情况下启动服务。隐式启动的代码如下。

```
final Intent serviceIntent=new Intent();
serviceIntent.setAction("cn.edu.neusoft.service.RandomService");
startService(serviceIntent);
```

显示启动则只需在 AndroidManifest.XML 文件中注册服务所在类。代码如下。

```
<service android:name=".RandomService"/>
```

在定义 Intent 时指明服务所在类，启动服务服务。代码如下。

```
final Intent serviceIntent=new Intent(this, RandomService.class);
startService(serviceIntent);
```

在绑定方式中，服务通过获取服务连接（Connection）实现。能够获取服务的对象实例，也就可以调用相应实现的函数，获取服务中的状态和数据信息。要使用服务的组件，可调用 bindService() 方法建立服务连接，调用 unbindService() 方法停止服务连接，并且一个服务可绑定多个服务连接，同时为多个组件提供服务。

通过绑定方式实现服务，需要重写 onBind() 方法，返回服务实例。

```
public class MyService extends Service {

private final IBinder iBinder=new LocalBinder();

    public class LocalBinder extends Binder {
MyServicegetService(){
return MyService.this;
}
    }
@Override
public IBinder onBind(Intent intent){
        return iBinder;
}
}
```

在 MainActivity 中定义 Intent 对象及绑定服务的方法，代码如下。

```
Intent serviceIntent=new Intent(MainActivity.this,MyService.class);
bindService(serviceIntent, serviceConnection,Context.BIND_AUTO_CREATE);
```

bindService()方法的第二个参数为服务连接对象，具体定义代码如下。

```
private ServiceConnection serviceConnection=new ServiceConnection(){
    @Override
    public void onServiceConnected(ComponentName name, IBinder service){
        MyService= ((MyService.LocalBinder)service).getService();
    }
    @Override
    public void onServiceDisconnected(ComponentName name){
        MyService=null;
    }
};
```

14.4　服务的生命周期

对于服务的生命周期，由于使用方式不同，经历的生命周期也不太一样。相比 Activity 的生命周期，服务执行的生命周期函数只是包含 onCreate()、onDestory()以及启动方式的 onStart()、绑定方式的 onBind()/onUnbind()。以启动方式实现服务，首先，第一次启动服务会执行 onCreate()→onStart()，直至调用 stop Service()时最后执行 onDestory()，如果服务已经运行，在整个生命周期过程中，onStart()会重复调用。而以绑定方式实现服务，首先执行 onCreate()→onBind()，而 onBind()将返回给客户端一个 IBind 接口实例，且允许客户端回调服务的方法，此时，调用者便和服务绑定成功。如果取消服务绑定，则调用 onUnbind()→onDestory()，重新绑定服务时，onRebind()将被调用。具体生命周期函数的调用次序如图 14.3 所示。

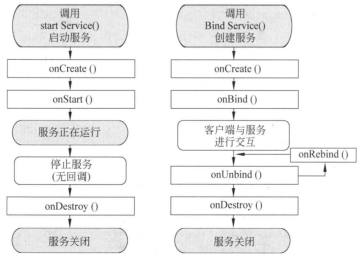

图 14.3　Service 生命周期

【例14-3】 随机数的加法运算。

本程序使用服务的特性进行两个随机数的加法运算,通过绑定方式使用服务,能够调用服务中的公有方法和属性,如果取消绑定,则调用不到服务的方法,无法进行加法运算。

在布局文件中,需要定义显示加法运算的 TextView 控件,以及绑定服务、取消绑定、加法运算的 3 个按钮,代码如下。

```xml
<?xml version="1.0" encoding="utf-8"?>
<LinearLayout xmlns:android="http://schemas.android.com/apk/res/android"
    android:layout_width="fill_parent"
    android:layout_height="fill_parent"
    android:orientation="vertical">
    <TextView
        android:id="@+id/info"
        android:layout_width="fill_parent"
        android:layout_height="wrap_content"
        android:gravity="center"></TextView>
    <TextView
        android:id="@+id/label"
        android:layout_width="fill_parent"
        android:layout_height="wrap_content"
        android:gravity="center"
        android:text="Service 示例"></TextView>
    <Button
        android:id="@+id/bind"
        android:layout_width="fill_parent"
        android:layout_height="wrap_content"
        android:text="服务绑定"></Button>
    <Button
        android:id="@+id/unbind"
        android:layout_width="fill_parent"
        android:layout_height="wrap_content"
        android:text="取消绑定"></Button>
    <Button
        android:id="@+id/compute"
        android:layout_width="fill_parent"
        android:layout_height="wrap_content"
        android:text="加法运算"></Button>
</LinearLayout>
```

在 MathService.java 中定义 Service,声明加法运算的 add()方法,完整代码如下。

```java
public class MathService extends Service {
    private final IBinder iBinder=new LocalBinder();
    private static String INFO="Service:加法运算";
    public class LocalBinder extends Binder {
```

```java
        MathService getService(){
            return MathService.this;
        }
    }

    @Override
    public IBinder onBind(Intent intent){
        Toast.makeText(this, "本地绑定:MathService", Toast.LENGTH_SHORT).show();
        return iBinder;
    }
    @Override
    public boolean onUnbind(Intent intent){
        Toast.makeText(this, "取消绑定:MathService", Toast.LENGTH_SHORT).show();
        return false;
    }
    public long add(long a, long b){
        return a+b;
    }
    public String showInfo(){
        return INFO;
    }
}
```

创建服务类后,在 MainActivity 类中以绑定方式实现服务,并调用服务类里的 add() 进行加法运算,完整代码如下。

```java
public class MainActivity extends Activity {
    private MathService mathService;
    private boolean isBound=false;
    private TextView resultLabel, info;
    private Button bind, unbind, compute;
    @Override
    public void onCreate(Bundle savedInstanceState){
        super.onCreate(savedInstanceState);
        setContentView(R.layout.activity_main);

        resultLabel= (TextView)findViewById(R.id.label);
        info= (TextView)findViewById(R.id.info);
        bind= (Button)findViewById(R.id.bind);
        unbind= (Button)findViewById(R.id.unbind);
        compute= (Button)findViewById(R.id.compute);

        bind.setOnClickListener(new View.OnClickListener(){
            @Override
            public void onClick(View v){
                if(!isBound){
                    Intent serviceIntent=new Intent(MainActivity.this,
```

```java
                        MathService.class);
                bindService(serviceIntent, iConnection,
                        Context.BIND_AUTO_CREATE);
                isBound=true;
            }
        }
    });
    unbind.setOnClickListener(new View.OnClickListener(){
        @Override
        public void onClick(View v){
            if(isBound){
                isBound=false;
                unbindService(iConnection);
                mathService=null;
            }
        }
    });
    compute.setOnClickListener(new View.OnClickListener(){
        @Override
        public void onClick(View v){
            if(mathService ==null){
                resultLabel.setText("未绑定服务,无法进行加法运算");
                return;
            }
            long a=Math.round(Math.random() * 100);
            long b=Math.round(Math.random() * 100);
            long result=mathService.add(a, b);
            String msg=String.valueOf(a)+"+"+String.valueOf(b)
                    +"="+String.valueOf(result);
            info.setText(mathService.showInfo());
            resultLabel.setText(msg);
        }
    });
}
private ServiceConnection iConnection=new ServiceConnection(){
    @Override
    public void onServiceConnected(ComponentName name, IBinder service){
        mathService= ((MathService.LocalBinder)service).getService();
    }
    @Override
    public void onServiceDisconnected(ComponentName name){
        mathService=null;
    }
};
}
```

运行程序前,需要在 Manifest 文件中定义 Service 标签,注册服务。代码如下。

```xml
<?xml version="1.0" encoding="utf-8"?>
<manifest xmlns:android="http://schemas.android.com/apk/res/android"
```

```
       package="cn.edu.neusoft.mathservice"
       android:versionCode="1"
       android:versionName="1.0">
    <uses-sdk
        android:minSdkVersion="14"
        android:targetSdkVersion="23" />
    <application
        android:allowBackup="true"
        android:icon="@drawable/ic_launcher"
        android:label="@string/app_name"
        android:theme="@style/AppTheme">
        <activity
            android:name="cn.edu.neusoft.mathservice.MainActivity"
            android:label="@string/app_name">
            <intent-filter>
                <action android:name="android.intent.action.MAIN" />
                <category android:name="android.intent.category.LAUNCHER" />
            </intent-filter>
        </activity>
        <service android:name=".MathService" />
    </application>
</manifest>
```

注册服务后，运行程序，单击"服务绑定"按钮，提示为本地绑定：MathService，表明服务已绑定成功。加法运算结果如图 14.4 所示。

如单击"取消绑定"按钮，提示为取消绑定：MathService，表明服务已取消绑定。进行加法运算，提示为未绑定服务，无法进行加法运算。运行结果如图 14.5 所示。

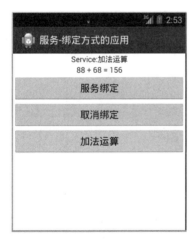

图 14.4　程序运行结果

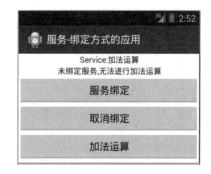

图 14.5　运行结果

本 章 小 结

本章介绍了广播的基本概念,通过广播思想实现生活中的应用场景,以及实现广播接收器的原理,运用广播接收器开发截获短信、显示来电位置等基本应用。介绍了服务的定义、应用场景以及生命周期。在实际应用中可以利用服务思想理解音乐播放器的逻辑,并以绑定方式实现服务进行加法运算。

本 章 习 题

1. 简述服务的基本原理与用途。
2. 简述服务的生命周期。
3. 简要概述使用服务实现音乐播放器的方法。
4. 简述广播的基本思想,如何基于广播思想设计一款 App?
5. 简述广播接收器的实现流程。
6. 通过对广播的理解,简要阐述系统发送广播消息的基本思路。
7. 通过 startService()和 bindService()启动服务,服务的生命周期上有哪些不同?

地图与定位

本章概述

本章讲解位置服务的概念,申请地图秘钥,通过百度地图来实现地图的显示、定位、导航等基本功能,以及百度地图的图层覆盖、位置标记的具体方法。

学习重点与难点

重点:

(1) 位置服务的概念。
(2) 地图的定位与导航功能。
(3) 申请百度地图秘钥。
(4) 设置与添加定位与导航功能的相关权限。

难点:

(1) 实现百度地图的定位与导航功能。
(2) 图层覆盖、位置标记的具体方法。

学习建议

掌握了位置服务的思想后,要实现百度地图的显示与定位功能,需在百度网站下载百度地图开发 SDK,并申请地图密钥。在 Android 应用程序中设置申请到的密钥及相关权限,根据具体的 API 文档实现定位、路线规划等接口方法,完成百度地图的显示、定位、路线规划等功能的开发。

15.1 位 置 服 务

位置服务(Location_Based Service,LBS),又称定位服务或基于位置的服务,融合了 GPS 定位、移动通信、导航等技术,提供与空间位置相关的综合应用服务。Android 平台支持提供位置服务的 API,在开发过程中,主要涉及 LocationManager 对象和 LocationProviders 对象。

LocationManager 可用来获取当前的位置,追踪设备的移动路线,设定敏感区域,在进入或离开敏感地区时,设备会发出指定警报。而 LocationProviders 提供定位功能的组件集合,集合中的每种组件以不同技术提供设备的当前位置,区别在定位的精度、速度、成本等方面。

15.2 地图的定义与显示

开发百度地图应用的第一步是在百度网站上申请"开发秘钥",然后使用 Android 系统提供的 MapView 控件显示和控制百度地图,同时可以在 MapView 上添加图层,实现地图表面的信息显示和图形绘制,通过对位置的定位,可实现从当前位置到目的地的路线规划,包括自驾路线、公交路线、步行路线,最后实现对所规划路线的导航功能。

15.2.1 申请地图密钥

为了在手机中更直观地显示地理信息,程序开发者可以直接应用百度提供的地图服务,显示百度地图。使用百度地图进行开发应用时,必须向百度申请经过验证的"地图密钥"(Map API Key),才能正常使用百度的地图服务。"密钥"是访问百度地图数据信息的密钥(API Key),无论是模拟器还是在真实设备中,都需要使用所申请的密钥。

申请"密钥"的第一步是注册一个百度账号,使用此账号展开接下来的密钥申请过程。注册百度账号的网址为 https://passport.baidu.com/,登录后页面如图 15.1 所示。

单击"立即注册"按钮进入百度账号注册界面,如图 15.2 所示。按要求填写相关注册信息,完成账号注册。

图 15.1 百度账号登录页面

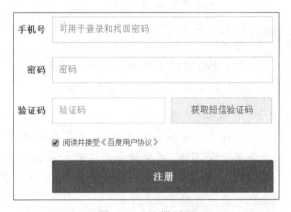

图 15.2 注册页面

使用已注册的账号登录百度地图开放平台(http://lbsyun.baidu.com/),单击右上角的控制台选项,进入密钥申请页面,如图 15.3 所示。

在页面中单击"创建应用"按钮,进入申请密钥的信息界面,如图 15.4 所示。

按要求填写信息,应用名称自行填写,应用类型选择 Android SDK,启用服务默认为全部选中。有关数字签名(SHA1),可单击页面中的查看详细配置信息,按文档的步骤获取数字签名,包名为开发百度地图的应用包名,请确认包名与所创建的项目包名完全一致。最后单击"提交"按钮,跳转到图 15.3 所示页面,应用名称即开发地图应用的名称,AK 码即为地图密钥,之后的开发中会用到此 AK(密钥)。

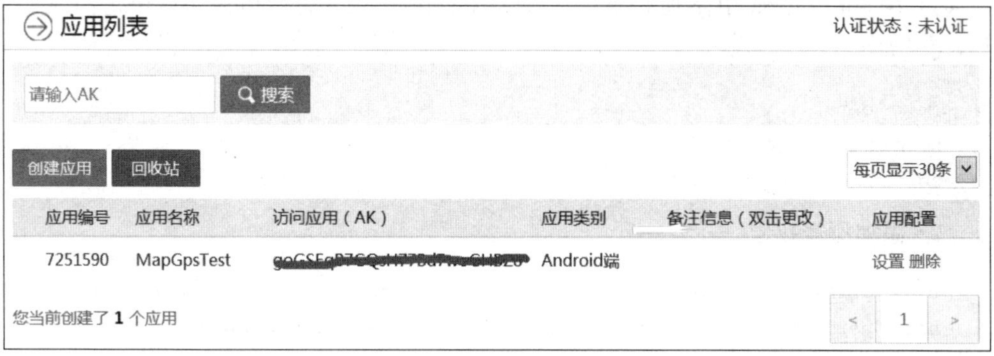

图 15.3　API 控制台页面

图 15.4　申请密钥页面

15.2.2　地图的显示

百度地图 SDK 为开发者提供了便捷的显示百度地图数据的接口，按照如下步骤操作，即可在应用中使用百度地图数据显示。

【例 15-1】　百度地图显示。

第一步：创建并配置工程；在 Android Studio 中创建要运行的项目。

第二步：在 AndroidManifest.xml 文件中添加开发密钥、所需权限等信息。

(1) 在 application 中添加开发密钥,将 android:value 属性值设为所申请的 AK(密钥),代码如下。

```xml
<meta-data
android:name="com.baidu.lbsapi.API_KEY"
android:value="Api Key" />
```

(2) 添加所需权限。

```xml
<uses-permission android:name="android.permission.ACCESS_NETWORK_STATE"/>
<uses-permission android:name="android.permission.INTERNET"/>
<uses-permission android:name="com.android.launcher.permission.READ_SETTINGS" />
<uses-permission android:name="android.permission.WAKE_LOCK"/>
<uses-permission android:name="android.permission.CHANGE_WIFI_STATE" />
<uses-permission android:name="android.permission.ACCESS_WIFI_STATE" />
<uses-permission android:name="android.permission.GET_TASKS" />
<uses-permission android:name="android.permission.WRITE_EXTERNAL_STORAGE"/>
<uses-permission android:name="android.permission.WRITE_SETTINGS" />
```

第三步,在布局 XML 文件中添加地图控件,代码如下。

```xml
<com.baidu.mapapi.map.MapView
    android:id="@+id/baiduMap"
    android:layout_width="fill_parent"
    android:layout_height="fill_parent"
    android:clickable="true" />
```

第四步,在应用程序创建时初始化 SDK 引用的 Context 全局变量,代码如下。

```java
public class MainActivity extends Activity {
    @Override
    protected void onCreate(Bundle savedInstanceState){
        super.onCreate(savedInstanceState);
        //在使用 SDK 各组件之前初始化 context 信息,传入 ApplicationContext
        //注意该方法要在 setContentView 方法之前实现
        SDKInitializer.initialize(getApplicationContext());
        setContentView(R.layout.activity_main);
    }
}
```

第五步,创建 Activity,管理地图的生命周期,代码如下。

```java
public class MainActivity extends Activity {
    private MapView mMapView=null;
        private BaiduMap=null;
    @Override
    protected void onCreate(Bundle savedInstanceState){
```

```
        super.onCreate(savedInstanceState);
        /**在使用SDK各组件之前初始化context信息,传入ApplicationContext
        注意该方法要在setContentView方法之前实现*/
        SDKInitializer.initialize(getApplicationContext());
        setContentView(R.layout.activity_main);
            //获取地图控件引用
        mMapView= (MapView)findViewById(R.id.baiduMap);
        baiduMap=mMapView.getMap();
        //普通地图
        baiduMap.setMapType(BaiduMap.MAP_TYPE_NORMAL);
        //卫星地图
        baiduMap.setMapType(BaiduMap.MAP_TYPE_SATELLITE);
      }
    @Override
    protected void onDestroy(){
        super.onDestroy();
        //执行onDestroy时执行mMapView.onDestroy(),实现地图生命周期管理
        mMapView.onDestroy();
    }
    @Override
    protected void onResume(){
        super.onResume();
        //执行onResume时执行mMapView. onResume(),实现地图生命周期管理
        mMapView.onResume();
    }
    @Override
    protected void onPause(){
        super.onPause();
        //执行onPause时执行mMapView. onPause(),实现地图生命周期管理
        mMapView.onPause();
    }
}
```

完成以上步骤后,运行程序,即可在模拟器或真机中正常显示百度地图,如图15.5所示。

百度地图的类型分为两种:普通地图和卫星地图。可使用setMapType()方法切换显示地图类型。代码如下。

```
baiduMap=mMapView.getMap();
//普通地图
baiduMap.setMapType(BaiduMap.MAP_TYPE_NORMAL);
//卫星地图
baiduMap.setMapType(BaiduMap.MAP_TYPE_SATELLITE);
```

效果如图15.6所示。

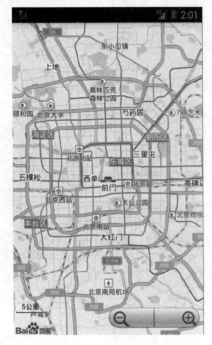
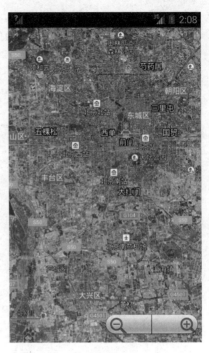

图 15.5　百度地图页面　　　　　　图 15.6　百度卫星地图

热力图是百度新上线的功能,用不同颜色的区块叠加在地图上,实时描述人群分布、密度和变化趋势,是基于百度大数据的一个便民出行服务。通俗地说,即显示地图上某一区域内人的密集程度。可通过 setBaiduHeatMapEnabled(true)方法实现,效果如图 15.7 所示。

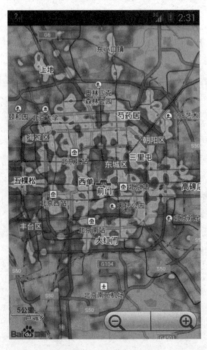

图 15.7　百度热力图

15.3 地图的定位及路线规划

15.3.1 定位原理

百度 SDK 定位，必须注册 GPS 和网络使用权限。百度地图采用 GPS、基站、WiFi 信号定位。当应用程序向定位 SDK 发起定位请求时，定位 SDK 会根据应用定位因素（GPS、基站、WiFi 信号）的实际情况（如是否开启 GPS、是否连接网络、是否有信号等）生成相应定位依据进行定位。具体定位如图 15.8 所示。

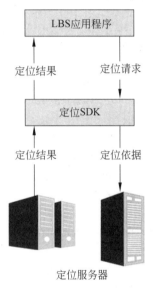

图 15.8　定位图示

如果设置 GPS 优先，则优先使用 GPS 定位；如果 GPS 定位未打开或没有可用位置信息，但网络连接正常，定位 SDK 则会返回网络定位（即 WiFi 与基站）的最优结果信息。为了使获得的网络定位结果更加精确，需打开手机的 WiFi 开关。百度地图提供的定位功能信息如图 15.9 所示

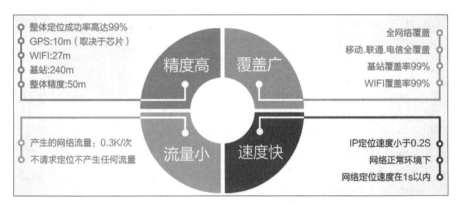

图 15.9　百度地图定位功能

15.3.2 定位与路线规划

实现百度地图的定位功能,需要下载最新的库文件以及 jar 文件,在百度地图开发平台(http://lbsyun.baidu.com)找到定位服务选项,下载 Android 定位 SDK 文件,将相应库文件复制到 libs/armeabi 目录下,将 jar 文件复制到新建工程的 libs 目录下,定位功能的相关 SDK 文件导入成功。

【例 15-2】 路线规划(百度地图)。

在布局文件(gps_activity.xml)中添加步行路线规划及定位请求的按钮,输入起点与终点的可编辑文本框,定义的内容如下。

```xml
<RelativeLayout xmlns:android="http://schemas.android.com/apk/res/android"
    xmlns:tools="http://schemas.android.com/tools"
    android:layout_width="match_parent"
    android:layout_height="match_parent"
    tools:context="cn.edu.neusoft.mapgpstest.GPSActivity" >
    <FrameLayout
        android:layout_width="match_parent"
        android:layout_height="match_parent" >
    <com.baidu.mapapi.map.MapView
        android:id="@+id/bmapView"
        android:layout_width="fill_parent"
        android:layout_height="fill_parent"
        android:clickable="true" />
    <LinearLayout
        android:layout_width="match_parent"
        android:layout_height="64dp"
        android:layout_gravity="top"
        android:background="@android:color/white"
        android:orientation="horizontal" >
    <EditText
        android:id="@+id/start_et"
        android:layout_width="wrap_content"
        android:layout_height="match_parent"
        android:layout_marginLeft="10dp"
        android:layout_weight="3"
        android:background="@null"
        android:focusable="true"
        android:focusableInTouchMode="true"
        android:hint="步行起点 "
        android:imeOptions="actionNext"
        android:inputType="text"
        android:maxLength="24"
```

```xml
            android:singleLine="true"
            android:textSize="14sp" />
        <EditText
            android:id="@+id/end_et"
            android:layout_width="wrap_content"
            android:layout_height="match_parent"
            android:layout_marginLeft="10dp"
            android:layout_weight="3"
            android:background="@null"
            android:hint="步行终点"
            android:imeOptions="actionNext"
            android:inputType="text"
            android:maxLength="24"
            android:singleLine="true"
            android:textSize="14sp" />
        <Button
            android:id="@+id/start_btn"
            android:layout_width="wrap_content"
            android:layout_height="40dp"
            android:layout_gravity="center_vertical"
            android:layout_marginRight="10dp"
            android:layout_weight="0.5"
            android:background="@drawable/btn_bg"
            android:gravity="center"
            android:text="步行路线"
            android:textColor="#FFFFFF"
            android:textSize="16sp" />
    </LinearLayout>
    <ImageView
        android:id="@+id/request_iv"
        android:layout_width="wrap_content"
        android:layout_height="wrap_content"
        android:layout_gravity="right|bottom"
        android:layout_marginBottom="20dp"
        android:layout_marginRight="20dp"
        android:scaleType="fitXY"
        android:src="@drawable/location" />
    </FrameLayout>
</RelativeLayout>
```

定位功能需要实现 BDLocationListener（位置监听器）接口，而路线规划要自定义路线规划结果监听器 OnGetRoutePlanResultListener，并重写相应方法。本项目只以步行路线为例,如想实现公交或驾车路线规划,在相应的重写方法中进行代码实现即可,Activity 的完整代码如下。

```java
package cn.edu.neusoft.mapgpstest;

public class GPSActivity extends Activity {
    private Toast toast;
    private MapView mapView;
    private BaiduMap baiduMap;
    /** 定位 */
    private LocationClient locationClient=null;
    private ImageView requestBtn;
    /** 路线规划对象 */
    private RoutePlanSearch rpSearch;
    /** 定位时中心点标注 */
    private Marker myLocationMarker;
    private String city="";
    private boolean isFirstLoc=true;
    private Button btnStart;
    private EditText etStart;
    private EditText etEnd;
    @Override
    protected void onCreate(Bundle savedInstanceState){
        super.onCreate(savedInstanceState);
        requestWindowFeature(Window.FEATURE_NO_TITLE);
        /**
         * 在使用 SDK 各组件之前初始化 context 信息,传入 ApplicationContext
         * 注意该方法要在 setContentView 方法之前实现
         */
        SDKInitializer.initialize(getApplicationContext());
        setContentView(R.layout.gps_activity);
        initView();
        baiduMap=mapView.getMap();
        locationClient=new LocationClient(getApplicationContext());
        etStart.requestFocus();
        LocationClientOption option=new LocationClientOption();
        /** 定位参数 */
        option.setOpenGps(true);              //打开 GPS
        option.setAddrType("all");            //返回的定位结果包含地址信息
        option.setCoorType("bd0911");         //返回的定位结果是百度经纬度,默认值 gcj02
        option.setLocationMode(LocationMode.Hight_Accuracy);
        /** 设置缩放控件不显示 */
        mapView.showZoomControls(false);
```

```java
            /** 设置地图参数 */
            locationClient.setLocOption(option);
            /** 注册定位监听器 */
            locationClient.registerLocationListener(new MyLocationListener());
            /** 启动定位 SDK */
            locationClient.start();
requestBtn.setOnClickListener(new OnClickListener(){
@Override
public void onClick(View v){
if(locationClient !=null &&locationClient.isStarted()){
                    showToast("正在定位……");
locationClient.requestLocation();
} else if(locationClient !=null){
locationClient.start();
} else {
                    Toast.makeText(GPSActivity.this, "LocationClient is null",
Toast.LENGTH_SHORT).show();
}
            }
        });
btnStart.setOnClickListener(new OnClickListener(){
@Override
public void onClick(View v){
                String start=etStart.getText().toString();
String end=etEnd.getText().toString();
routePlan(start, end, city);
}
        });
}
private void showToast(String msg){
if(toast ==null){
toast=Toast.makeText(this, msg, Toast.LENGTH_SHORT);
} else {
toast.setText(msg);
toast.setDuration(Toast.LENGTH_SHORT);
}
toast.show();
}
/**
     * 初始化获取地图控件引用
     */
public void initView(){
mapView= (MapView)findViewById(R.id.bmapView);
```

```java
requestBtn=(ImageView)findViewById(R.id.request_iv);
btnStart=(Button)findViewById(R.id.start_btn);
etStart=(EditText)findViewById(R.id.start_et);
etEnd=(EditText)findViewById(R.id.end_et);
}
/**
     * 将当前位置移动到指定位置
     *
public void updateLocationPosition(BaiduMap eBikeAMap, LatLng cenpt,
                                   float zoom){
/* 定义地图状态 */
MapStatus mMapStatus=new MapStatus.Builder().target(cenpt).zoom(zoom)
            .build();
/* 定义 MapStatusUpdate 对象,以便描述地图状态将要发生的变化 */
MapStatusUpdate mMapStatusUpdate=MapStatusUpdateFactory
            .newMapStatus(mMapStatus);
/* 改变地图状态 */
baiduMap.setMapStatus(mMapStatusUpdate);
}
/**
     * 路线规划
     */
public void routePlan(String start, String end, String city){
rpSearch=RoutePlanSearch.newInstance();
rpSearch.setOnGetRoutePlanResultListener(MyGetRoutePlanResultListener);
/** 起点与终点 */
PlanNode stNode=PlanNode.withCityNameAndPlaceName(city, start);
PlanNode enNode=PlanNode.withCityNameAndPlaceName(city, end);
/** 步行路线规划 */
boolean res=rpSearch.walkingSearch(new WalkingRoutePlanOption().from(
            stNode).to(enNode));
}
/**
     * 定位监听器
     */
public class MyLocationListener implements BDLocationListener {
@Override
public void onReceiveLocation(BDLocation location){
if(location ==null)
return;
LatLng cenpt=new LatLng(location.getLatitude(),
location.getLongitude());
        try {
```

```java
//以定位的城市为默认
try {
                String addr=location.getAddrStr();
city=addr.substring(addr.indexOf("省")+1,
addr.indexOf("市")+1);
} catch(Exception e){
            }
/** 定义 Marker 坐标点 */
LatLng point=new LatLng(cenpt.latitude, cenpt.longitude);
//构建 Marker 图标
BitmapDescriptor bitmap=BitmapDescriptorFactory
                    .fromResource(R.drawable.real_time_location);
//构建 MarkerOption,用于在地图上添加 Marker
OverlayOptions option=new MarkerOptions().position(point)
                    .icon(bitmap);
//在地图上添加 Marker,并显示
myLocationMarker=(Marker)baiduMap.addOverlay(option);
isFirstLoc=false;
Log.d("Log", "==>经度: "+cenpt.longitude+" 纬度: "
+cenpt.latitude);
updateLocationPosition(baiduMap, cenpt, 18);
} catch(Exception e){
            e.printStackTrace();
}
        }
    }
/**
    * 路线规划结果监听
    */
OnGetRoutePlanResultListener MyGetRoutePlanResultListener=new
OnGetRoutePlanResultListener(){
/**步行*/
@Override
public void onGetWalkingRouteResult(WalkingRouteResult routeResult){
//TODO 步行路线规划
if(routeResult ==null
|| routeResult.error !=SearchResult.ERRORNO.NO_ERROR){
            Toast.makeText(GPSActivity.this, "抱歉,未找到结果", Toast.LENGTH_
            SHORT)
                    .show();
}
if(routeResult.error ==SearchResult.ERRORNO.AMBIGUOUS_ROURE_ADDR){
/** 起终点或途经点地址有歧义,通过以下接口获取建议查询信息 */
```

```
              routeResult.getSuggestAddrInfo();
                    return;
          }
          if(routeResult.error ==SearchResult.ERRORNO.NO_ERROR){
                    WalkingRouteOverlay overlay=new MyWalkingRouteOverlay(
          baiduMap);
          baiduMap.setOnMarkerClickListener(overlay);
          overlay.setData(routeResult.getRouteLines().get(0));
          overlay.addToMap();
          overlay.zoomToSpan();
          }
                    }
          /** 获取驾车线路规划结果 */
          @Override
          public void onGetDrivingRouteResult(DrivingRouteResult arg0){
          //TODO 驾车路线规划
          }
          /** 获取公交换乘路径规划结果 */
          @Override
          public void onGetTransitRouteResult(TransitRouteResult arg0){
          //TODO 公交路线规划
          }
              };
          /**
               * 继承步行规划的子类,通过覆盖相应方法实现功能
               */
          class MyWalkingRouteOverlay extends WalkingRouteOverlay {
          public MyWalkingRouteOverlay(BaiduMap arg0){
          super(arg0);
          }
                }
          @Override
          protected void onDestroy(){
          super.onDestroy();
          mapView.onDestroy();
                  if(rpSearch !=null){
          rpSearch.destroy();
          }
                }
          @Override
          protected void onResume(){
          super.onResume();
          mapView.onResume();
```

```
    }
    @Override
    protected void onPause(){
    super.onPause();
    mapView.onPause();
    }
    }
```

最后，在 AndroidManifest.XML 文件中配置地图密钥、权限、服务信息，将＜meta-data＞中的 android：value 属性值设为所申请的 AK(密钥)。

```
<meta-data
android:name="com.baidu.lbsapi.API_KEY"
android:value="API KEY" />

<service
    android:name="com.baidu.location.f"
    android:enabled="true"
    android:process=":remote" >
</service>
<uses-permission android:name="android.permission.WRITE_SETTINGS" />
<!--这个权限用于进行网络定位 -->
<uses-permission android:name="android.permission.ACCESS_COARSE_LOCATION" />
<!--这个权限用于访问GPS定位 -->
<uses-permission android:name="android.permission.ACCESS_FINE_LOCATION" />
<!--用于访问WiFi网络信息,WiFi信息会用于进行网络定位 -->
<uses-permission android:name="android.permission.ACCESS_WIFI_STATE" />
<!--获取运营商信息,用于支持提供运营商信息相关的接口 -->
<uses-permission android:name="android.permission.ACCESS_NETWORK_STATE" />
<!--这个权限用于获取WiFi的获取权限,WiFi信息会用来进行网络定位 -->
<uses-permission android:name="android.permission.CHANGE_WIFI_STATE" />
<!--用于读取手机当前的状态 -->
<uses-permission android:name="android.permission.READ_PHONE_STATE" />
<!--写入扩展存储,向扩展卡写入数据,用于写入离线定位数据 -->
<uses-permission android:name="android.permission.WRITE_EXTERNAL_STORAGE" />
<!--访问网络,网络定位需要上网 -->
<uses-permission android:name="android.permission.INTERNET" />
<!--SD卡读取权限,用户写入离线定位数据 -->
<uses-permission android:name="android.permission.MOUNT_UNMOUNT_FILESYSTEMS"
/>
<!--允许应用读取低级别的系统日志文件 -->
<uses-permission android:name="android.permission.READ_LOGS" />
```

建议用真机执行程序，提高定位的准确性，运行结果如图 15.10 所示。

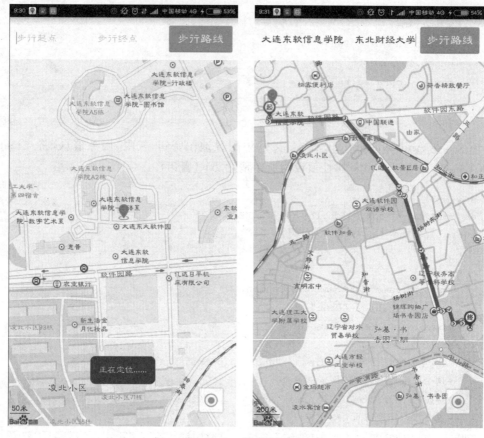

图 15.10　路线规划图

本章小结

　　本章介绍了位置服务(Location_Based Service)的概念,使用百度地图 SDK 实现地图的显示,申请百度地图密钥的流程,以基于对位置服务的学习进一步完成百度地图的定位功能,对定位后的位置进行标记,并以步行路线规划为例,实现路线规划的功能。

本章习题

　　1. 讨论位置服务和地图应用的发展前景。
　　2. 如何下载百度地图 SDK、申请百度地图密钥?
　　3. 简述百度地图定位与路线规划的实现思想。
　　4. 简述地图路线规划中的覆盖图层如何。当开发一款地图导航的移动应用时,如何先实现定位功能,再通过定位的位置实现路线规划,最后根据规划的路线进行实时导航?

参 考 文 献

[1] 李宁. Android 开发权威指南[M]. 北京：人民邮电出版社,2013.
[2] 王向辉,张国印,赖明珠. Android 应用程序开发[M]. 北京：清华大学出版社,2012.
[3] 明日科技. Android 从入门到精通[M]. 北京：清华大学出版社,2015.
[4] 陈会安. Android SDK 程序设计与开发范例[M]. 北京：清华大学出版社.2013.
[5] 刘甫迎. Android 移动编程实用教程[M]. 北京：电子工业出版社,2012.
[6] Bill Philips Brain Hardy. Android 编程权威指南[M]. 王明发,译. 北京：人民邮电出版社,2014.
[7] Android(Google 公司开发的操作系统)[EB/OL]. [2017-10-10]. http://baike.baidu.com.